AF682168

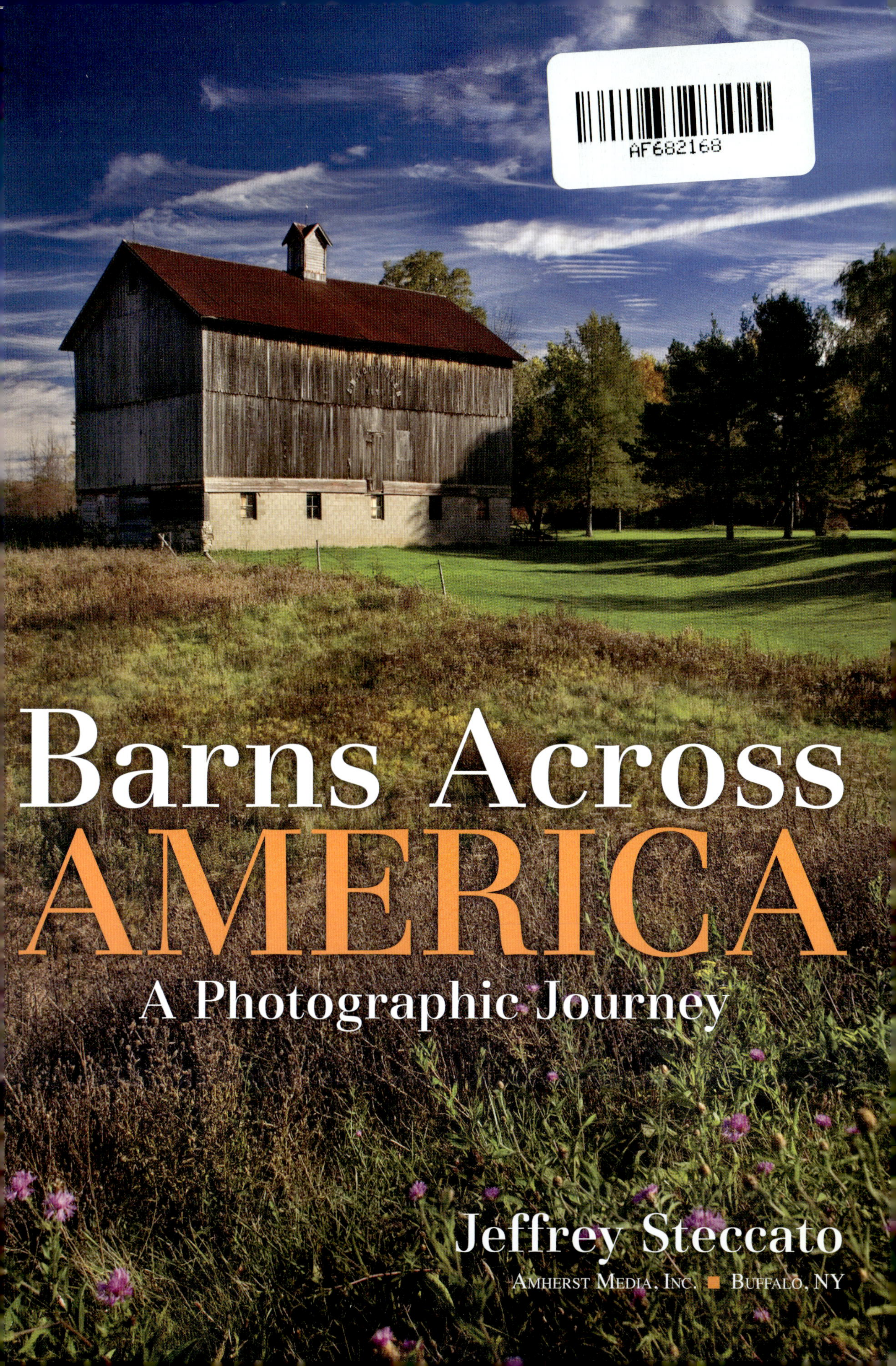

# Barns Across AMERICA

## A Photographic Journey

Jeffrey Steccato

Amherst Media, Inc. ■ Buffalo, NY

# Dedication

To my wife of 30 years, Susan, and our three children: Jamie, Jeffrey, and Thomas. Thank you for all of the support and patience you've given me when, during family vacations, I would pull over and spend an hour photographing American barn icons. Thank you for making this book happen. I love you all.

Copyright © 2019 by Jeffrey Steccato.
All rights reserved.
All photographs by the author unless otherwise noted.

Published by:
Amherst Media, Inc.
PO BOX 538
Buffalo, NY 14213
www.AmherstMedia.com

Publisher: Craig Alesse
Associate Publisher: Katie Kiss
Senior Editor/Production Manager: Michelle Perkins
Editors: Barbara A. Lynch-Johnt, Beth Alesse
Acquisitions Editor: Harvey Goldstein
Editorial Assistance from: Carey A. Miller, Roy Bakos, Jen Sexton-Riley, Rebecca Rudell
Business Manager: Sarah Loder
Marketing Associate: Tonya Flickinger

ISBN-13: 978-1-68203-412-5

Library of Congress Control Number: 2019932342

Printed in the United States of America
10 9 8 7 6 5 4 3 2 1

No part of this publication may be reproduced, stored, or transmitted in any form or by any means, electronic, mechanical, photocopied, recorded or otherwise, without prior written consent from the publisher.

Notice of Disclaimer: The information contained in this book is based on the author's experience and opinions. The author and publisher will not be held liable for the use or misuse of the information in this book.

**AUTHOR A BOOK WITH AMHERST MEDIA**

Are you an accomplished photographer with devoted fans? Consider authoring a book with us and share your quality images and wisdom with your fans. It's a great way to build your business and brand through a high-quality, full-color printed book sold worldwide. Our experienced team makes it easy and rewarding for each book sold—no cost to you. E-mail **submissions@amherstmedia.com** *today.*

www.facebook.com/AmherstMediaInc
www.youtube.com/AmherstMedia
www.twitter.com/AmherstMedia
www.instagram.com/amherstmediaphotobooks

# Contents



**below**—A barn adorned in red, white, and blue; it's pure Americana. Wyoming.

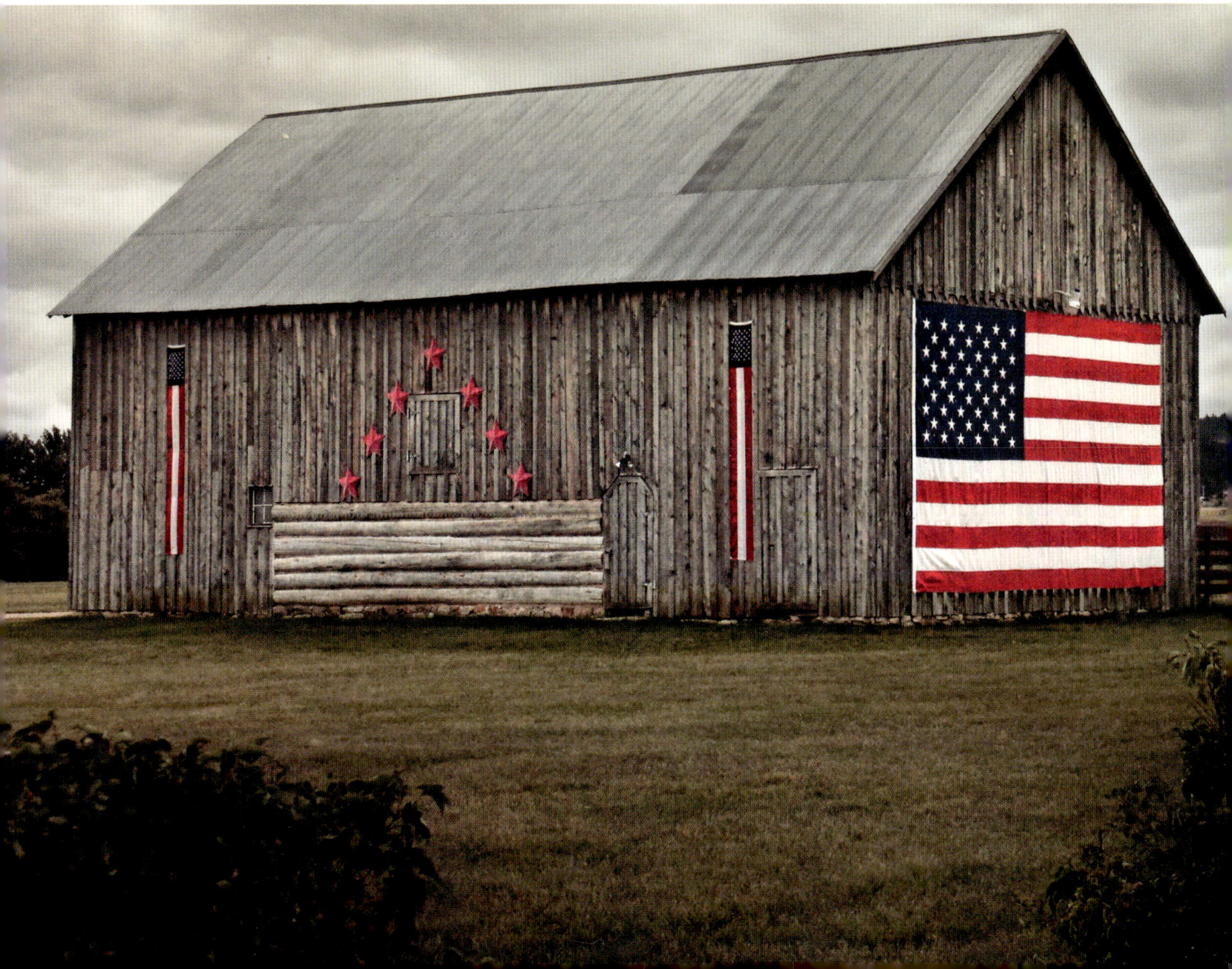

# About the Author

Jeffrey's photographic journey began in high school, when he began taking pictures for local newspapers, covering high school sporting events. It was then he met master photographer Carmen Schettino, who opened the door for Jeffrey, teaching him black & white darkroom techniques, as well as commercial/corporate photography.

In 1982, Jeffrey became the apprentice of master photographer Theodore Haviland, of Haviland Photography, in Ho-Ho-Kus, New Jersey. He worked with children and families, which heightened his desire to become a professional photographer.

After many years of being mentored and learning the "old school" tradition of printmaking, Jeffrey purchased Haviland Photography upon his mentor's 1997 retirement.

Throughout his career as a professional photographer, Jeffrey has entered several of his photographs in over 100 professional photography competitions and placed in every one of them. He has also been the recipient of many awards, such as the distinguished Courts of Honor Award, Judges Choice Awards, and prestigious Kodak Gallery Awards. His work has appeared on magazine covers and calendars, and has also been displayed in the Professional Photographers Association coveted Loan Collection books several times throughout his career.

In August 2000, after obtaining his certification, Jeffrey also achieved his master's degree in photography. In April, 2010, he received the Professional Photographers National Award for his meritorious contributions to his profession. These awards are the highest honor bestowed by the Professional Photographers Association, ranking Jeffrey in the top 3 percent of photographers nationwide. In 2012, Jeffery received The Humanitarian of the Year Award from the Hermitage Museum Foundation, located in Ho-Ho-Kus, New Jersey, for the time and talent he has so graciously given to this foundation.

In recent years, Jeffrey has been most fortunate to exhibit his works of art in various local galleries, including the historic Hermitage Museum. This opportunity has provided him the chance to express and display his heart's desire to preserve a part of America's history through his image collection of ghost towns and various barns in the United States.

Jeffrey's goal is to share his vision of the world, as he has experienced it from the perspective of his camera lens throughout his 35 years as a professional photographer. He strives to create images that have a perfect balance of light and composition, evoke true emotion, and tell a story.

# Acknowledgments

To my early mentors, who inspired me to pick up a camera and start my career as a professional photographer: Bill Savage, my high school photography teacher/wrestling coach taught me the basics of photography. Carmen Schettino, master photographer, saw a teenager with the potential to become a photographer. Theodore Z. Haviland, master photographer, took 21-year-old me into his world of photography and made me his apprentice, teaching me the fundamentals of family portrait photography and the everyday skills of running a photography business. After 38 years, I still take great pride in keeping the Haviland Photography name and tradition alive. 2019 marks our 70th anniversary.

**below**—A time-worn barn, surrounded by yellow weeds. Kentucky.

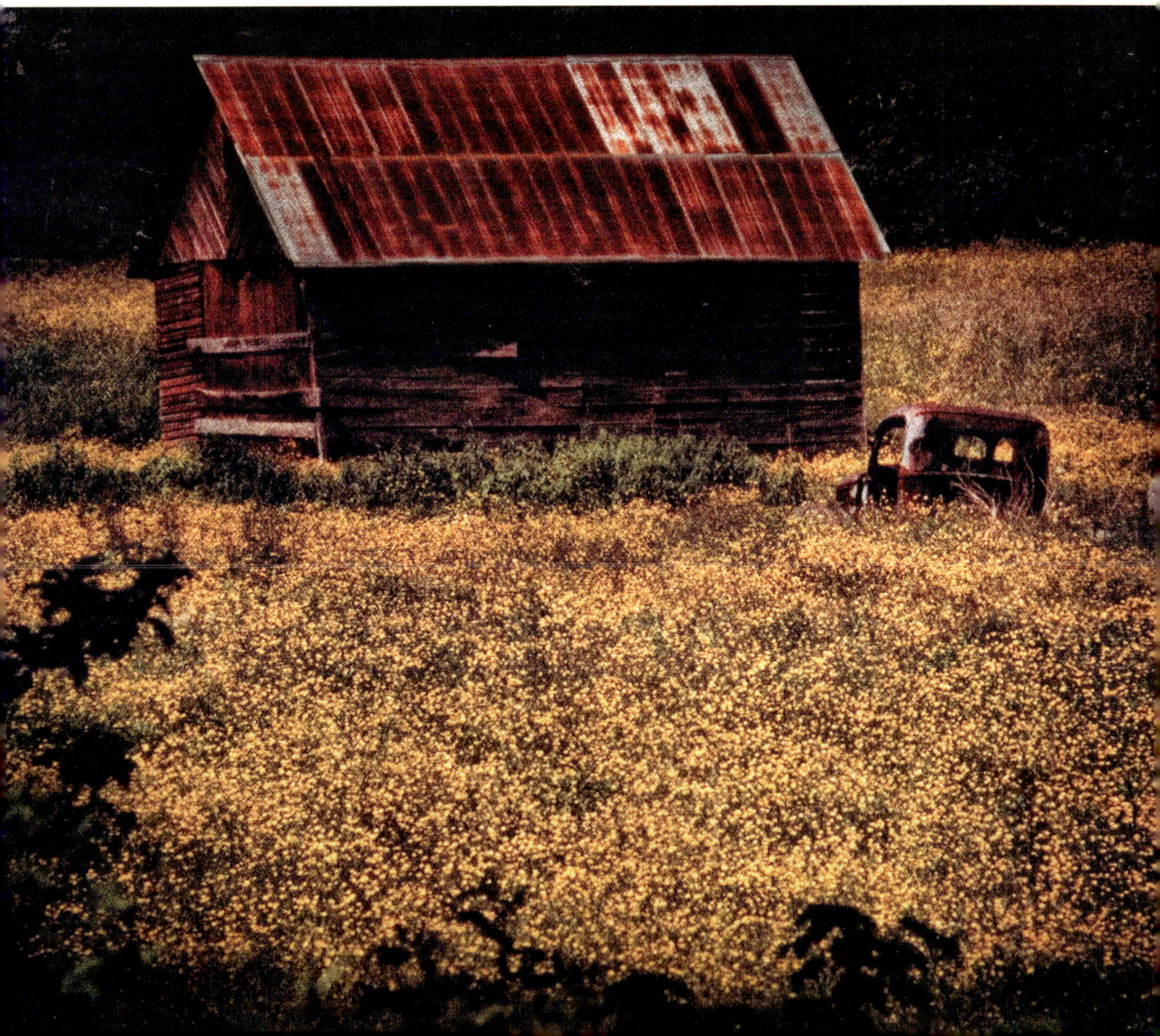

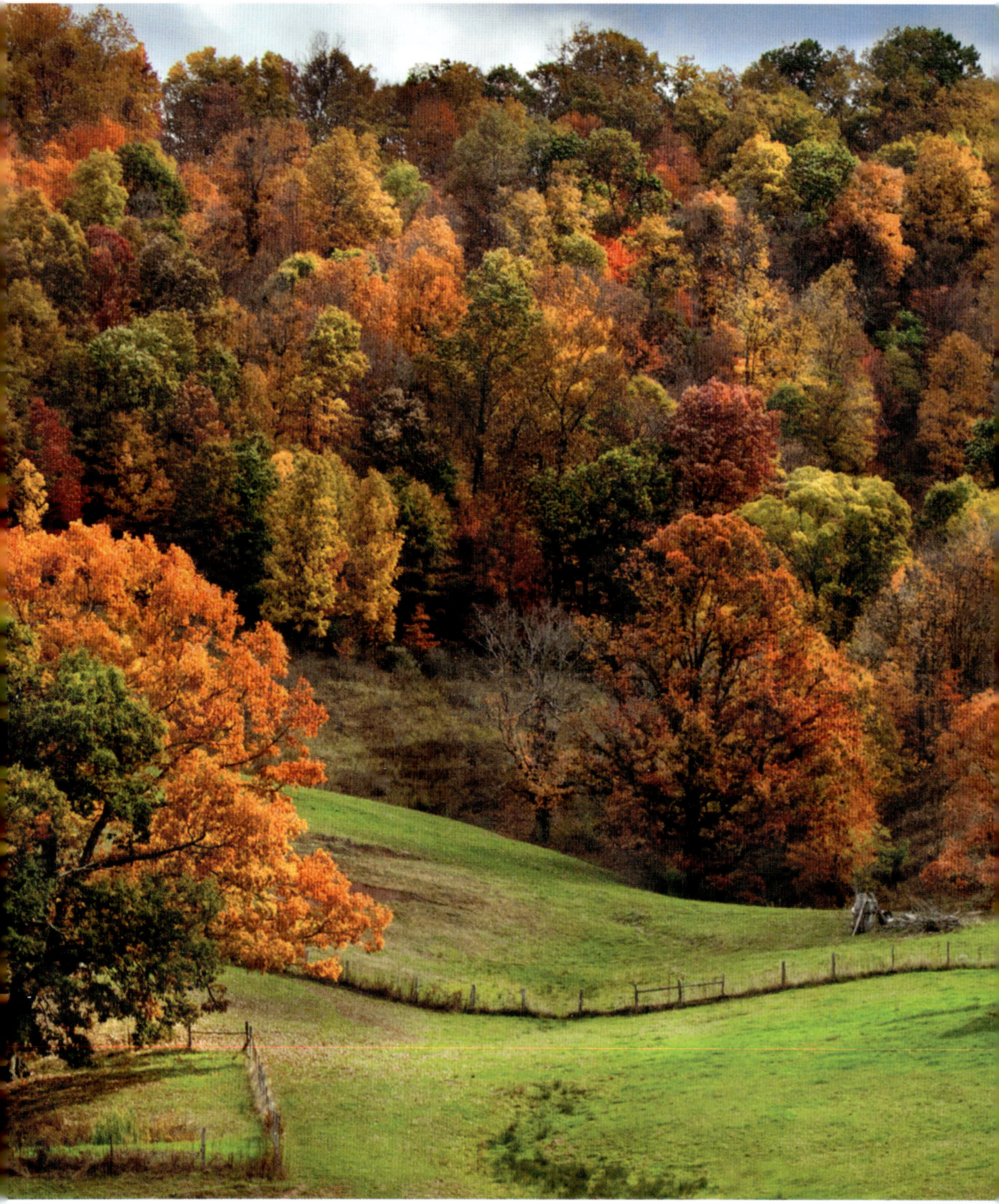

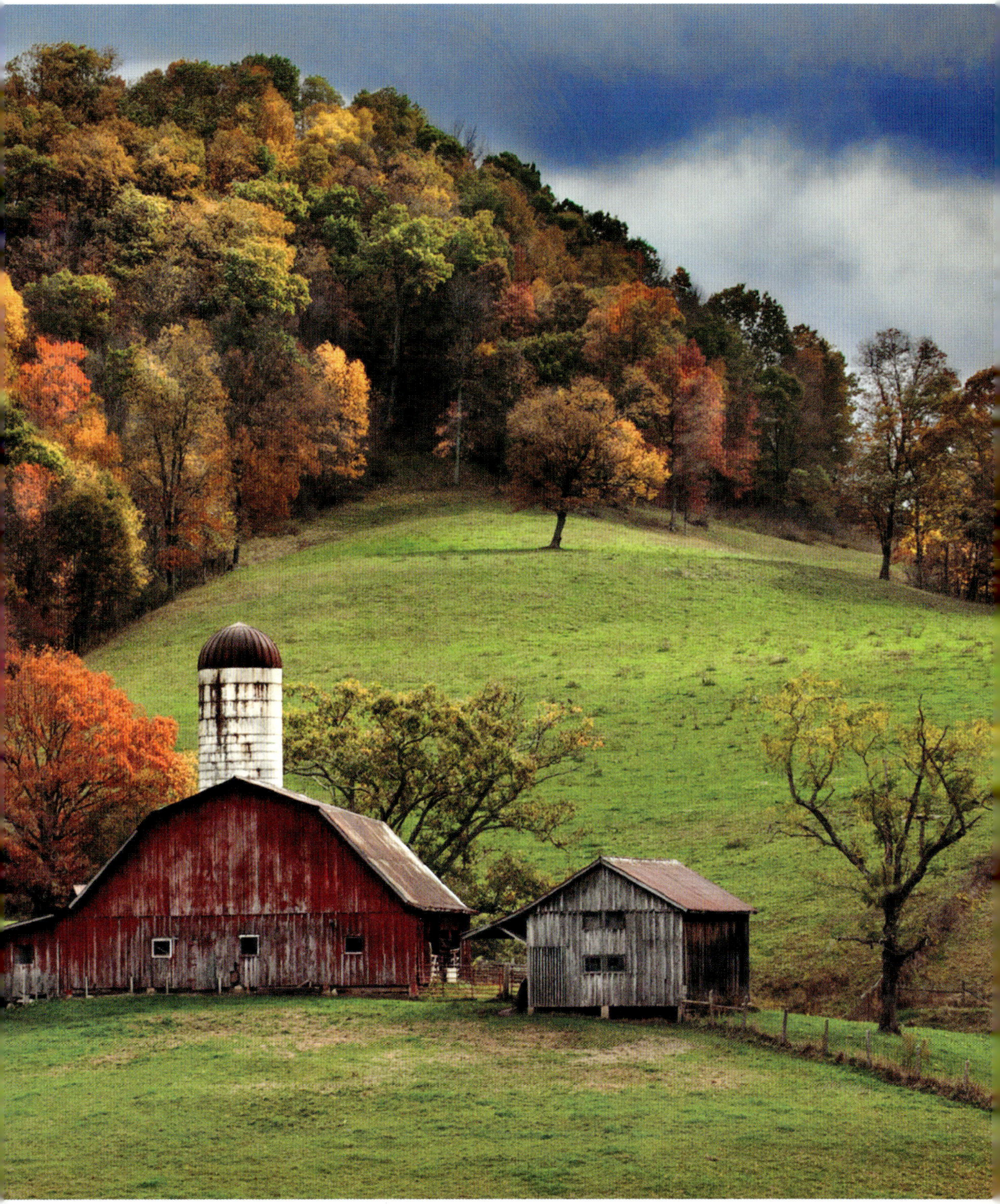

**above—***Autumn's Radiance*. West Virginia.

# The American Barn

In 1935, 6.8 million farms operated across the United States, most with at least one barn on their property. Now, just around 650,000 barns remain standing.

Barns are a threatened part of our heritage. We need to capture their beauty, document their history, and encourage preservation.

There are all kinds of threats to the barns that dot the American landscape—including weather, fire, and age—but the biggest obstacle they face is not having a purpose.

## A Bit of History

The English word "barn" comes from two middle-English words: "bereb," which means barley, and "aern," which signifies "place." Together, they mean "a place for barley."

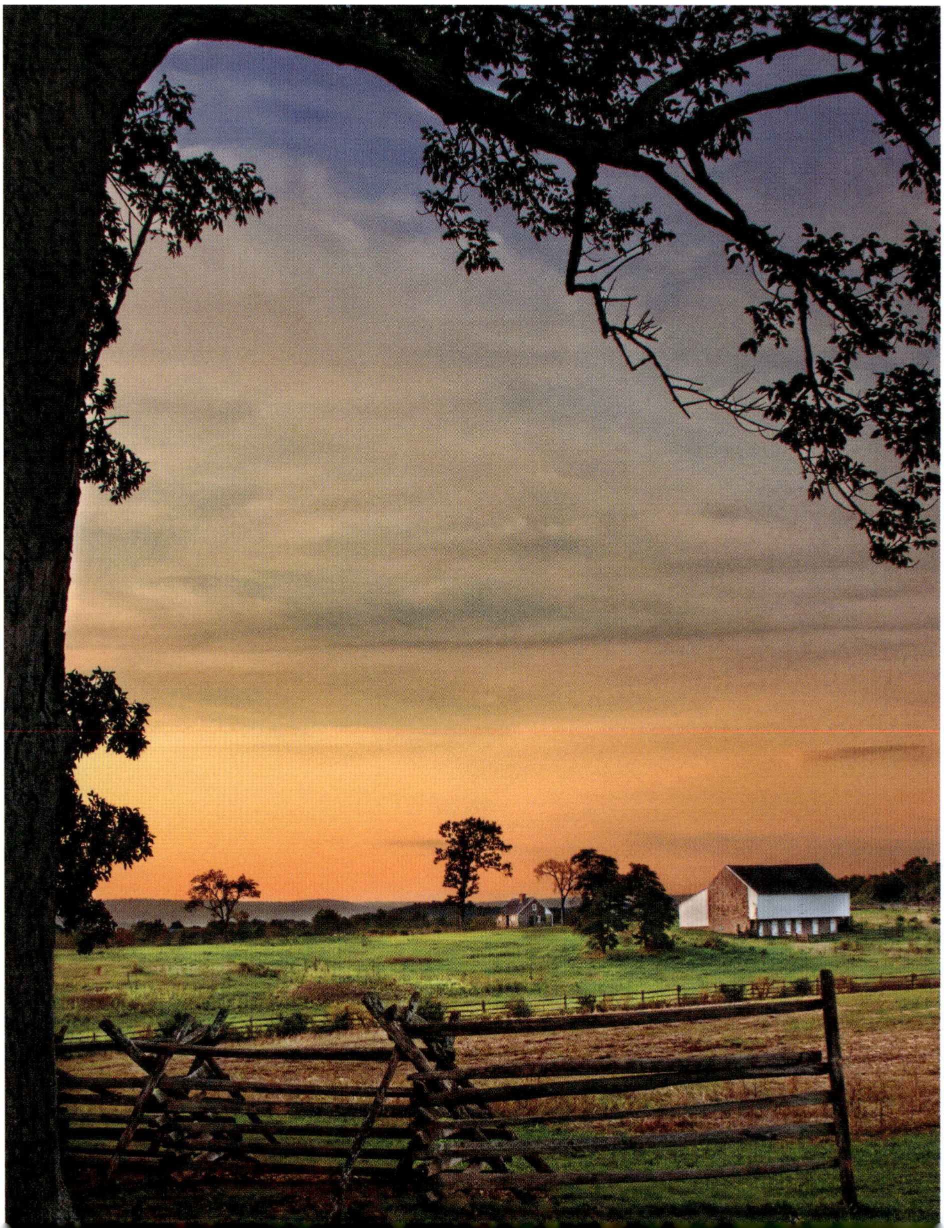

**left—**
*At Peace.*
Gettysburg, Pennsylvania.

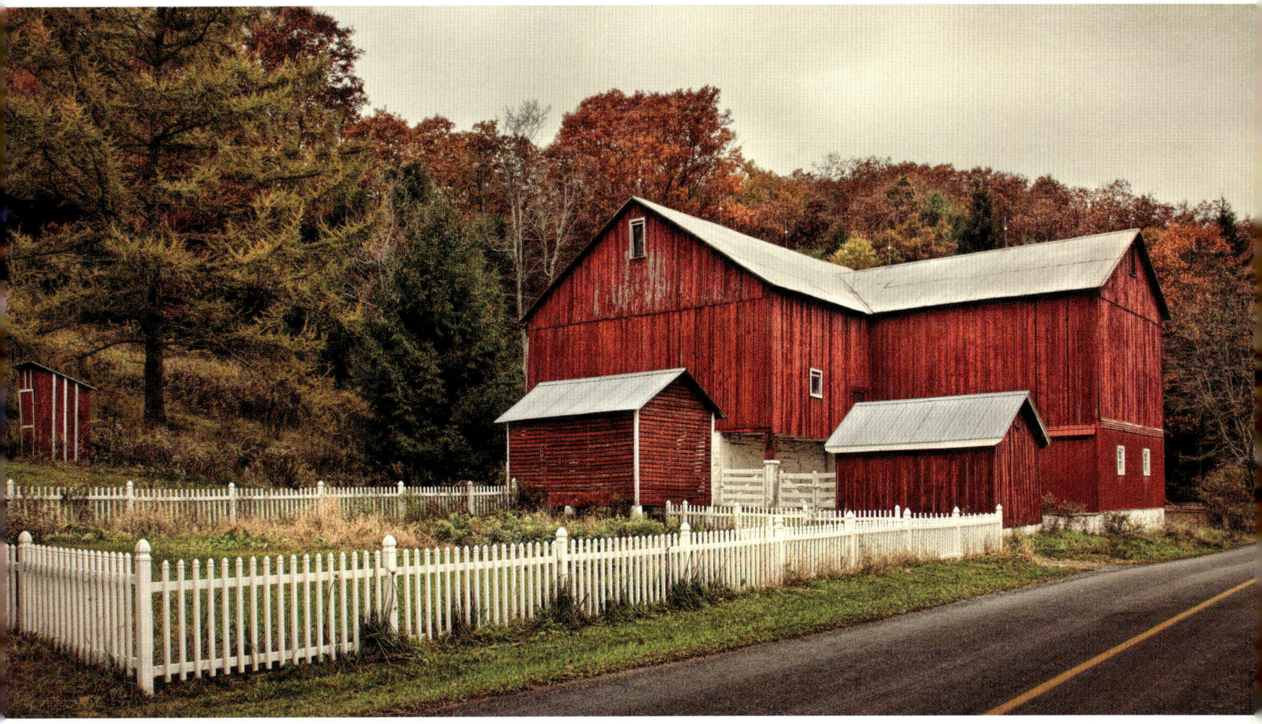

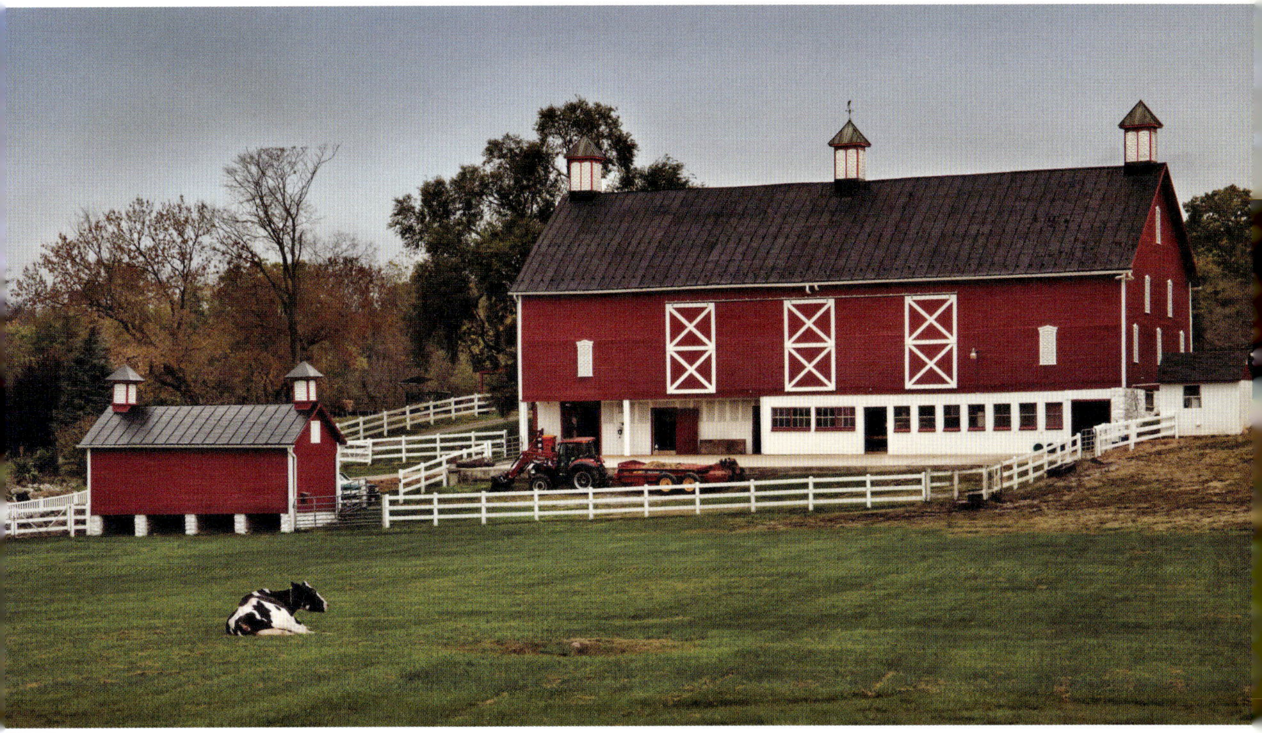

**top**—An American dream, white picket fence and all. Pennsylvania.

**bottom**—A well-cared for, operational barn located in rural Maryland.

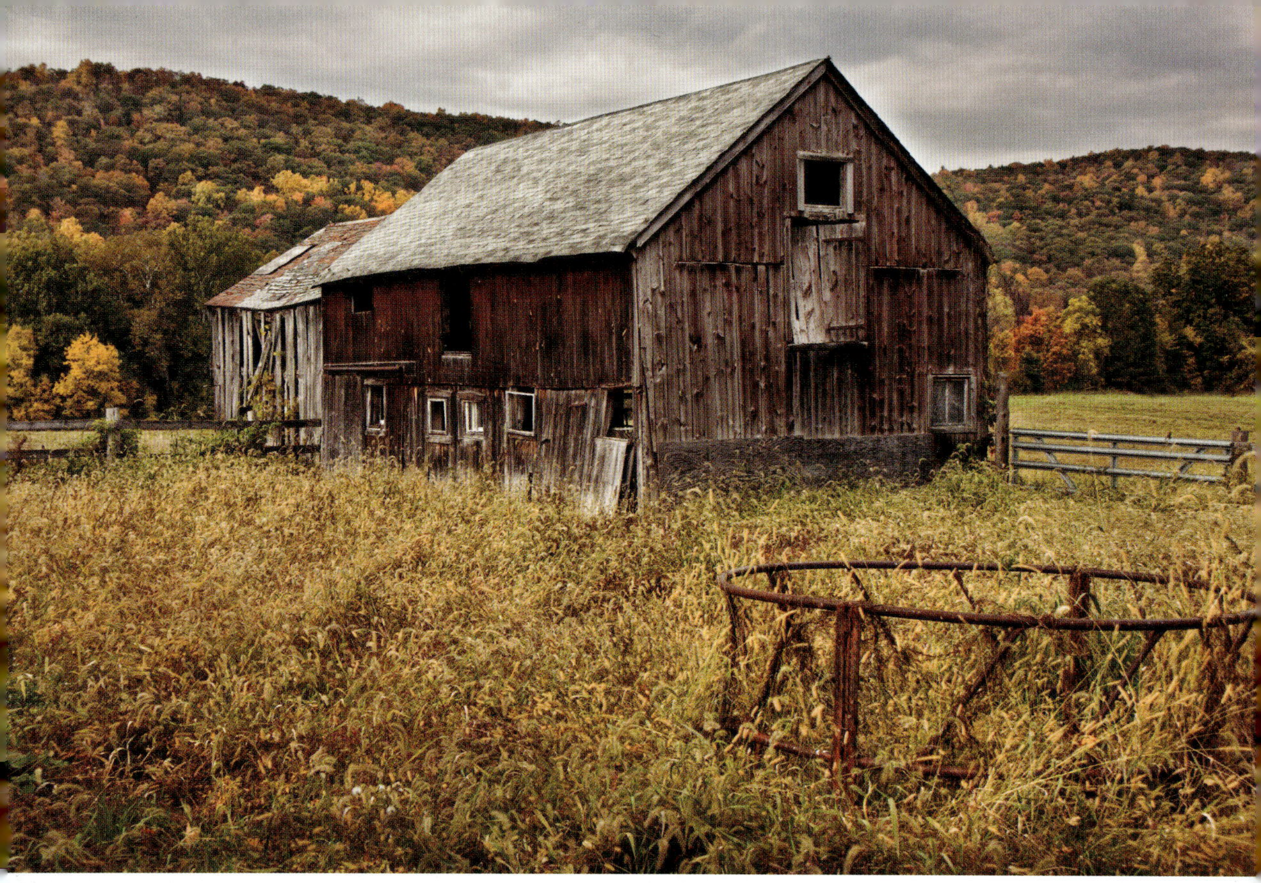

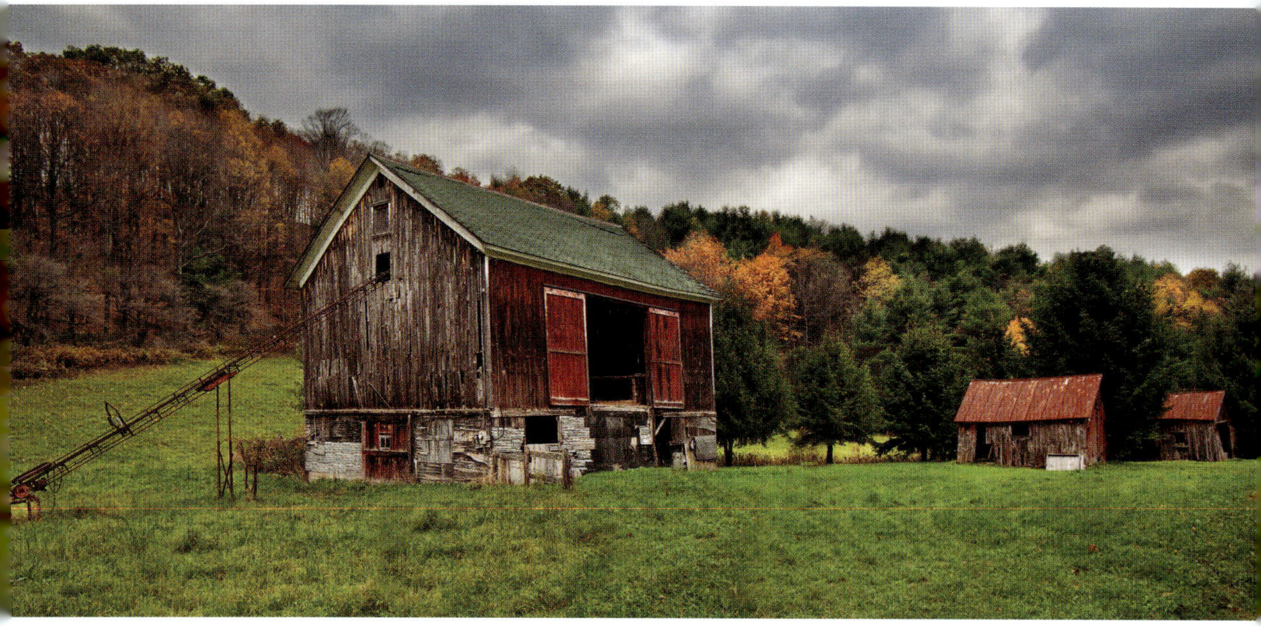

**top**—Connecticut.

**bottom**—New York.

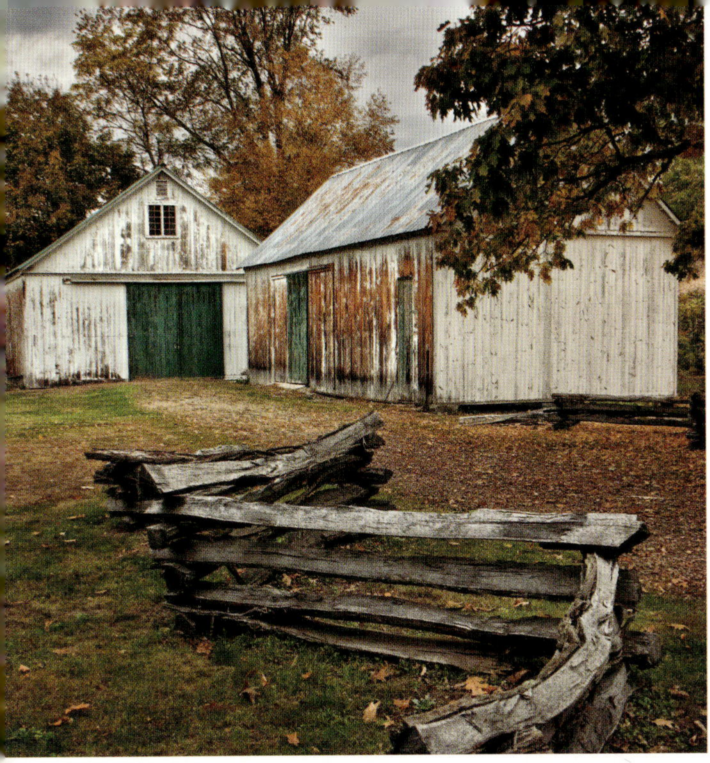

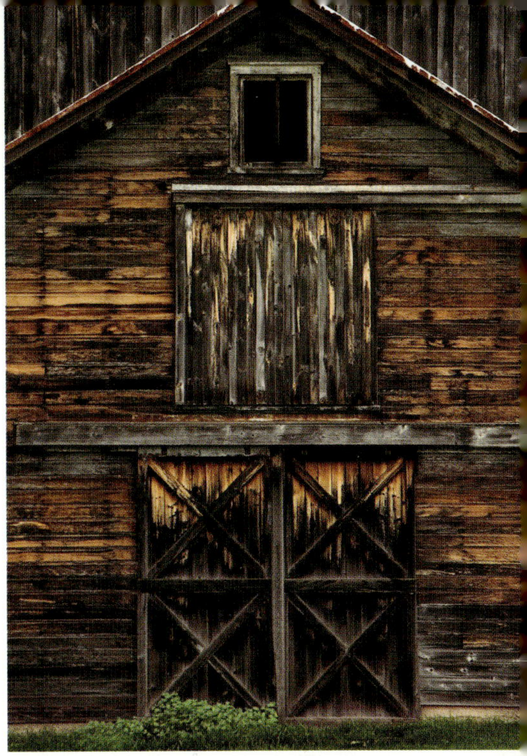

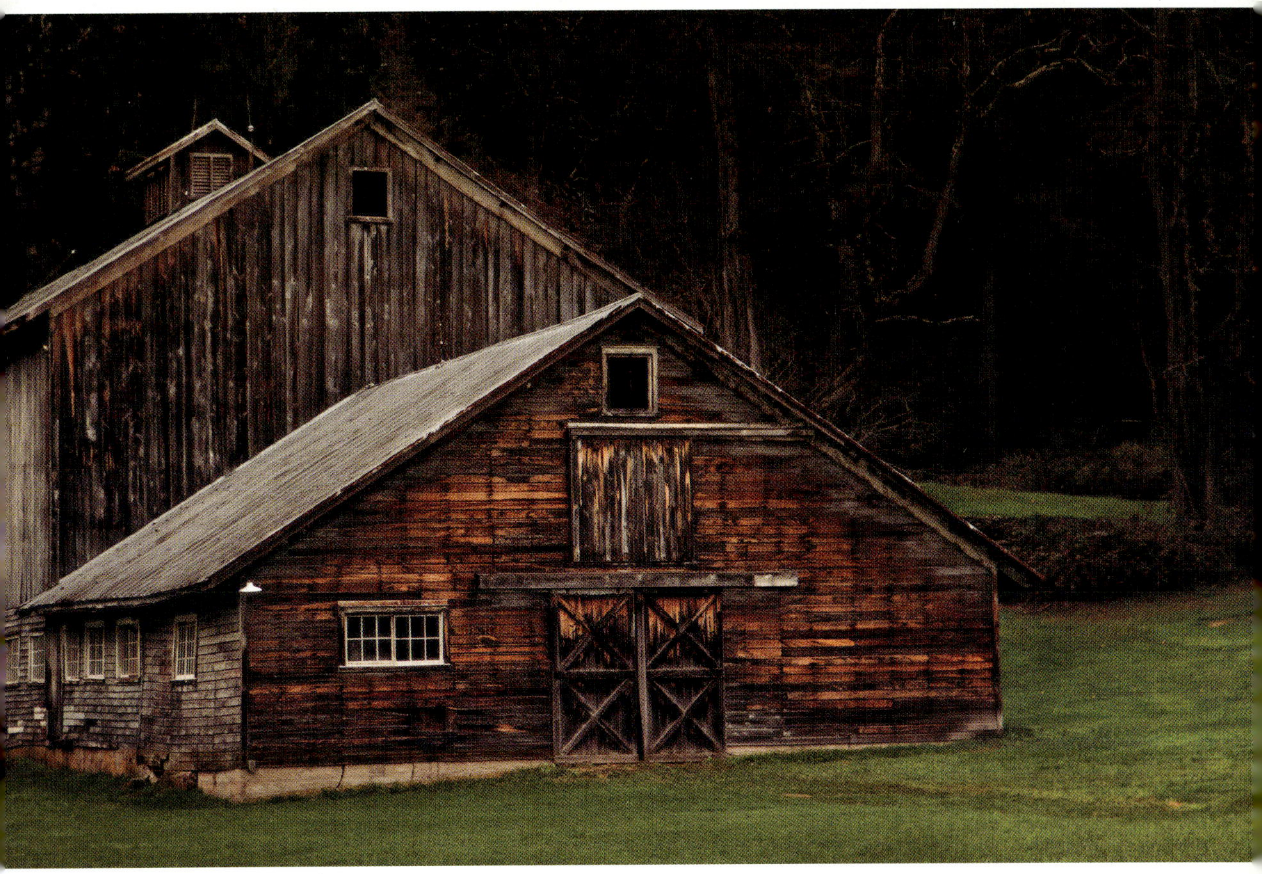

**top left**—New Hampshire.

**top right**—*Weathered Through Time*. New York.

**above**—New York.

# Bank Barns

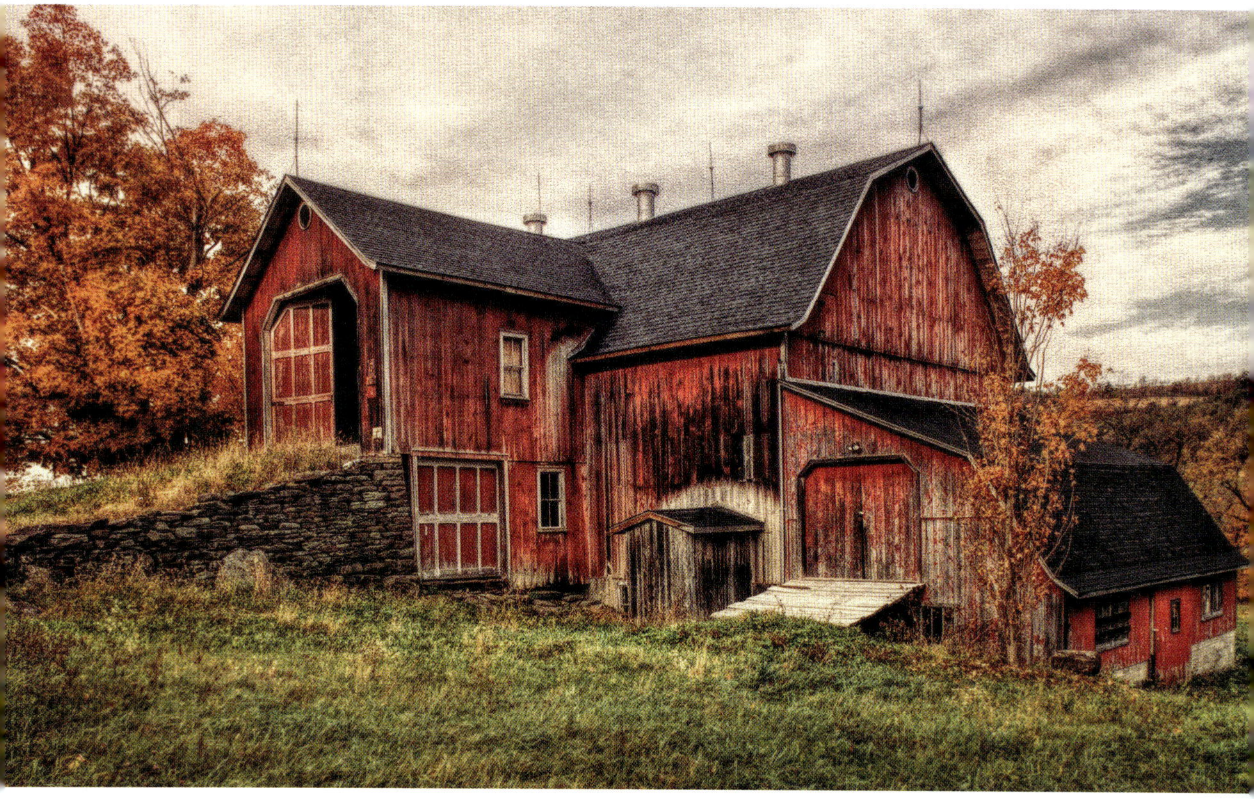

A bank barn, or banked barn, is noted for its accessibility at ground level. These barns are often built into the side of a hill, or a bank. Both the upper and lower floors are accessible from the ground level, with one door at the top of the hill, and another at the bottom.

A related barn style is the high-bridgeway barn. Though similar to bank barns, high-bridgeway barns are different in that if a hill was not available, a ramp/bridgeway was built to allow entry into the barn.

Occasionally, you may run across another variation of the high-bridgeway barn: one with its entrance at the gable end, rather than on the side.

**above**—This high-bridgeway barn was photographed on film in 1985. New York.

**following page, top and bottom**—Two high-bridgeway barns. One has fallen into disrepair; the other has been fully restored. Northern New York State.

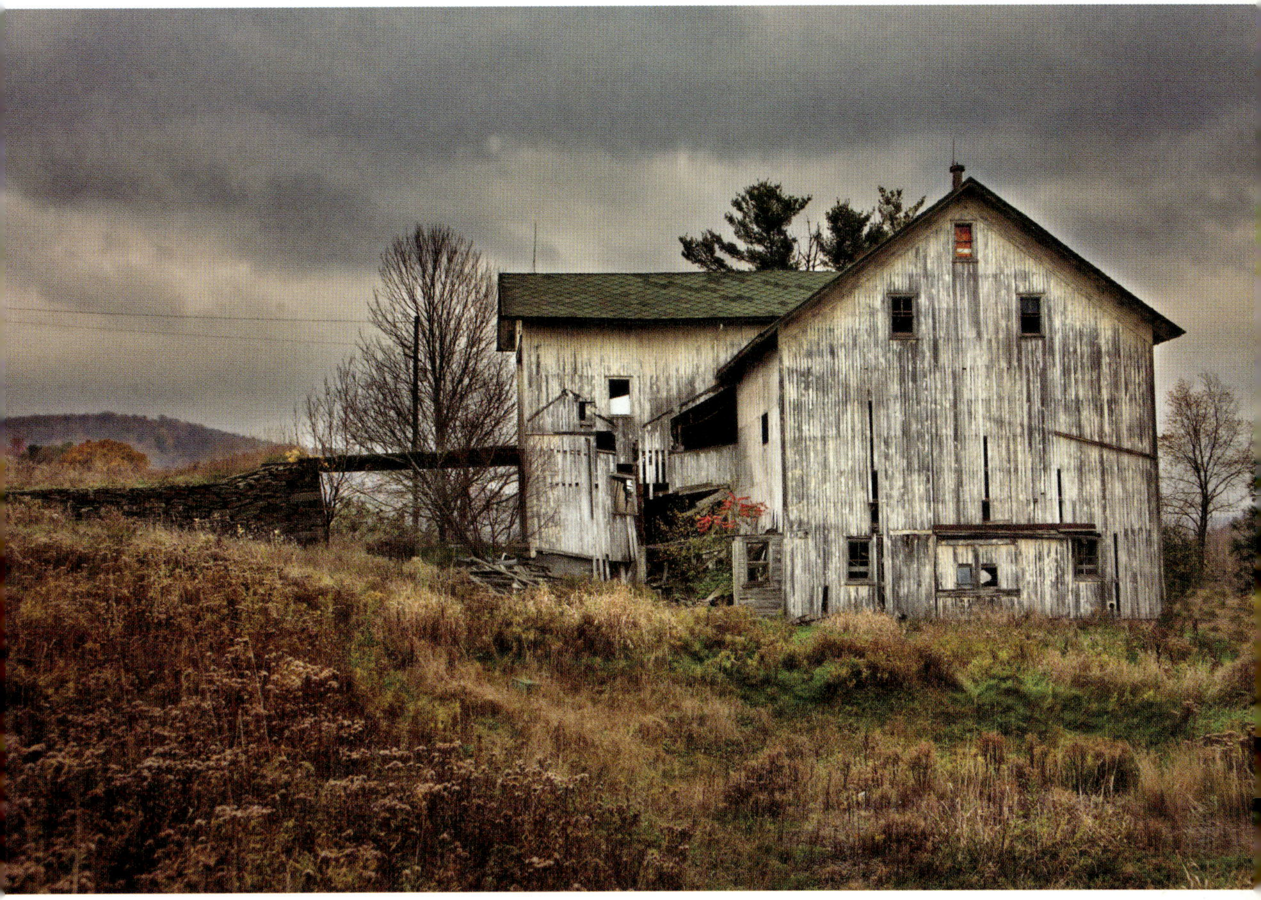

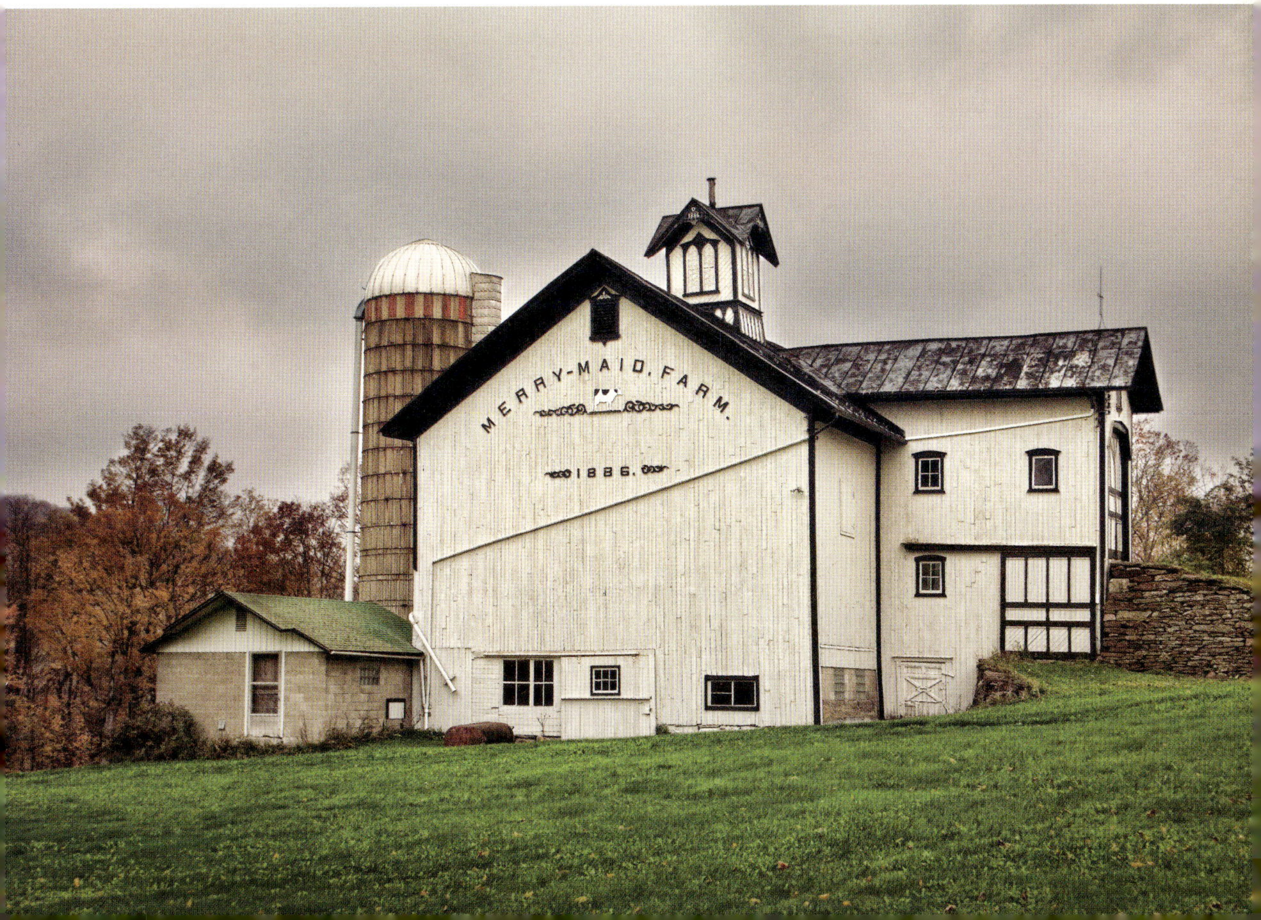
MERRY-MAID FARM
1886

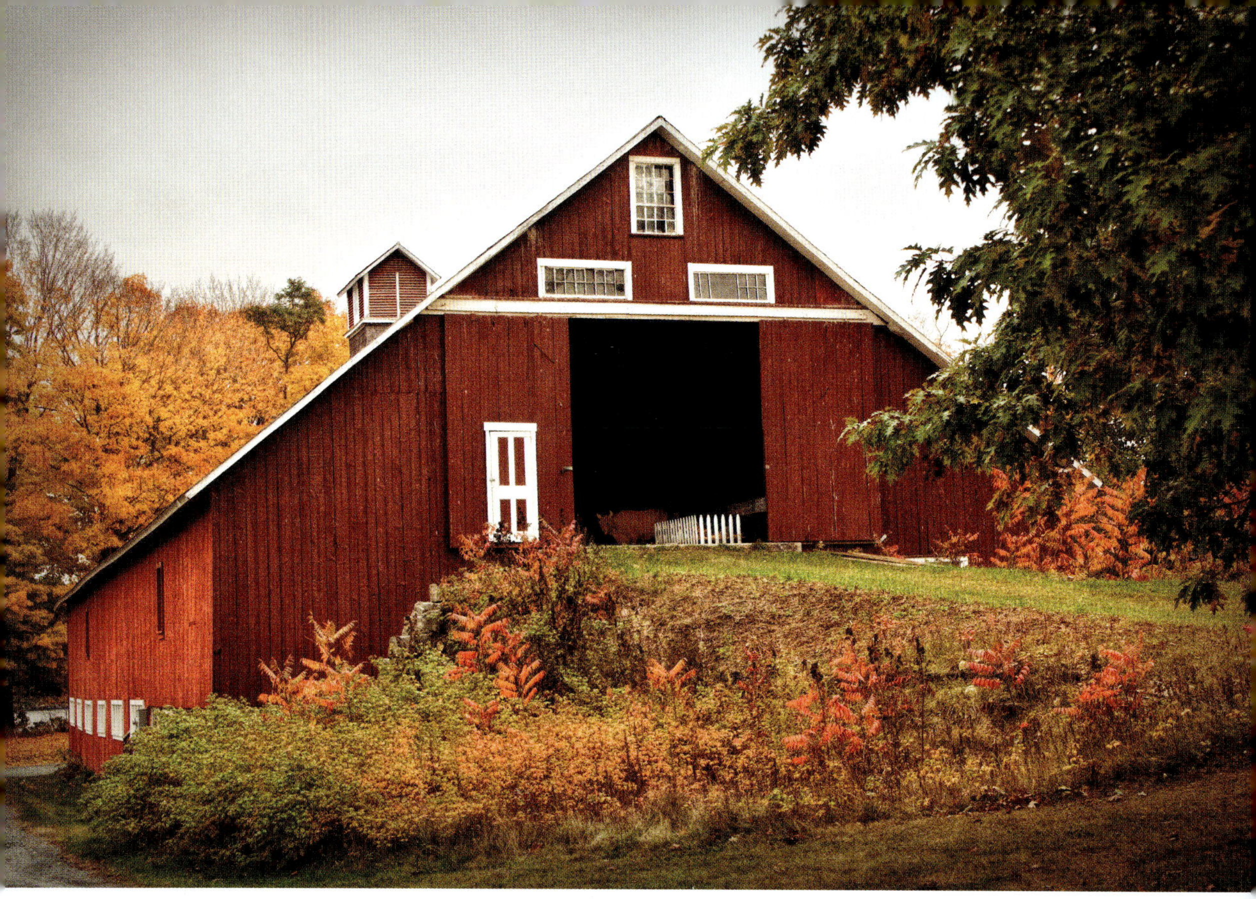

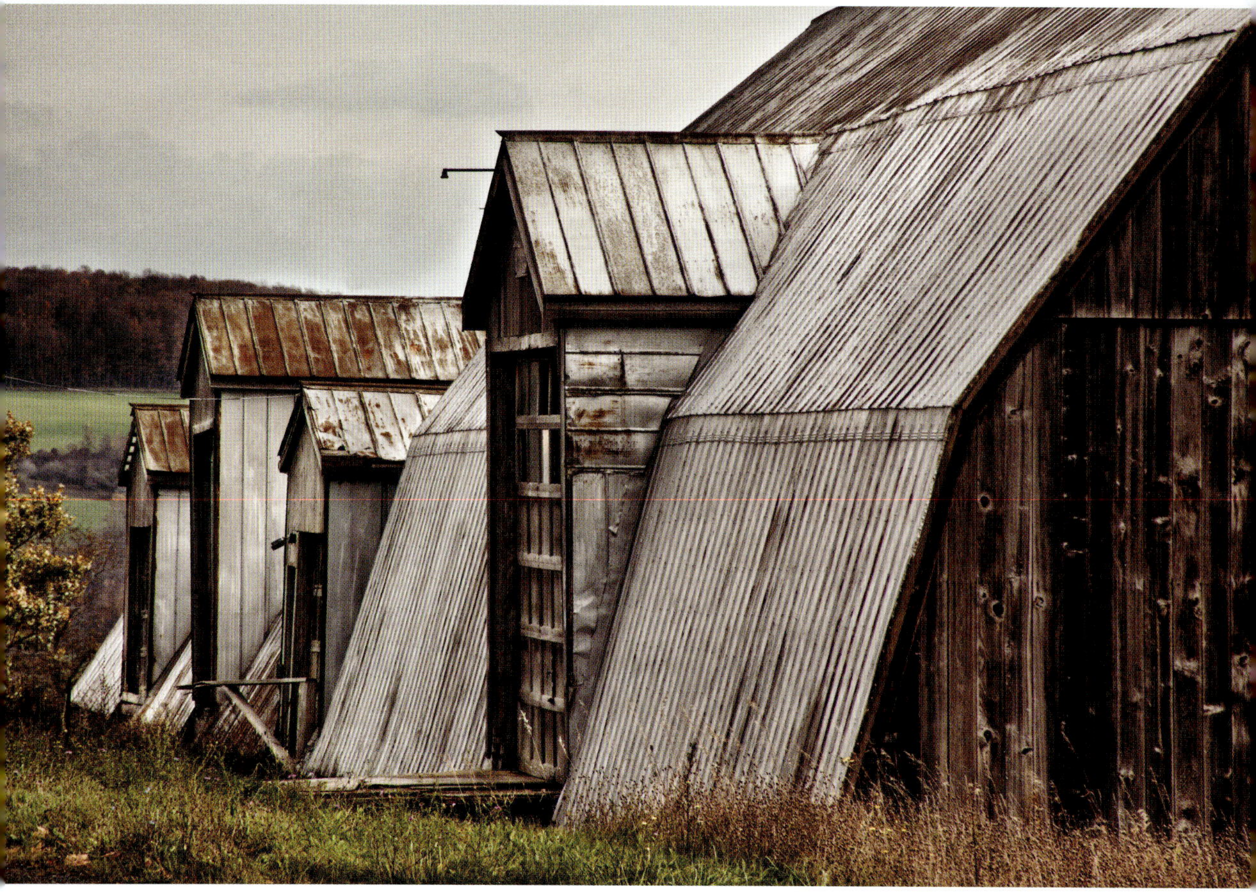

**previous page, top**—Gable-entry high-bridgeway barn. New Hampshire.

**previous page, bottom**—Double-entry high-bridgeway dairy barn. Two barns next to each other and attached, flowing from one to the other.

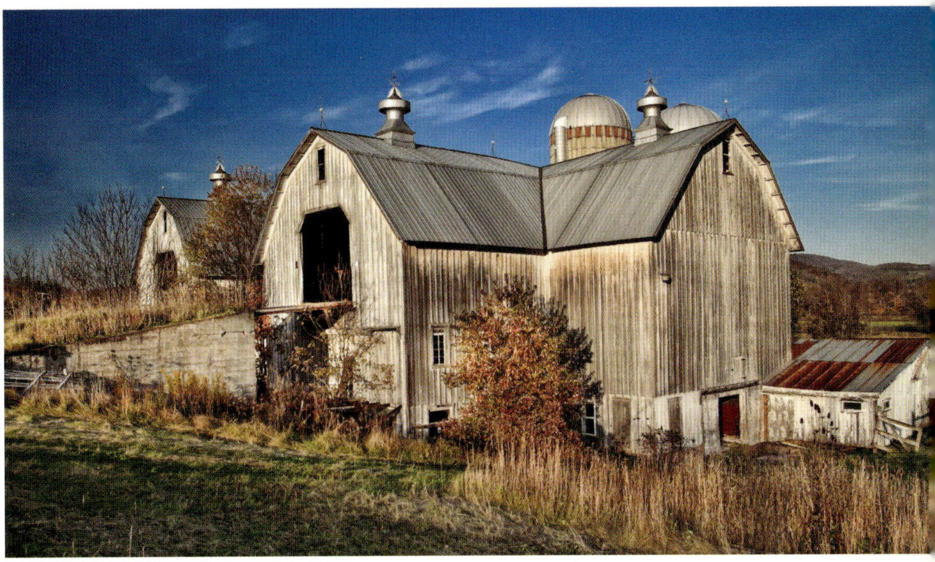

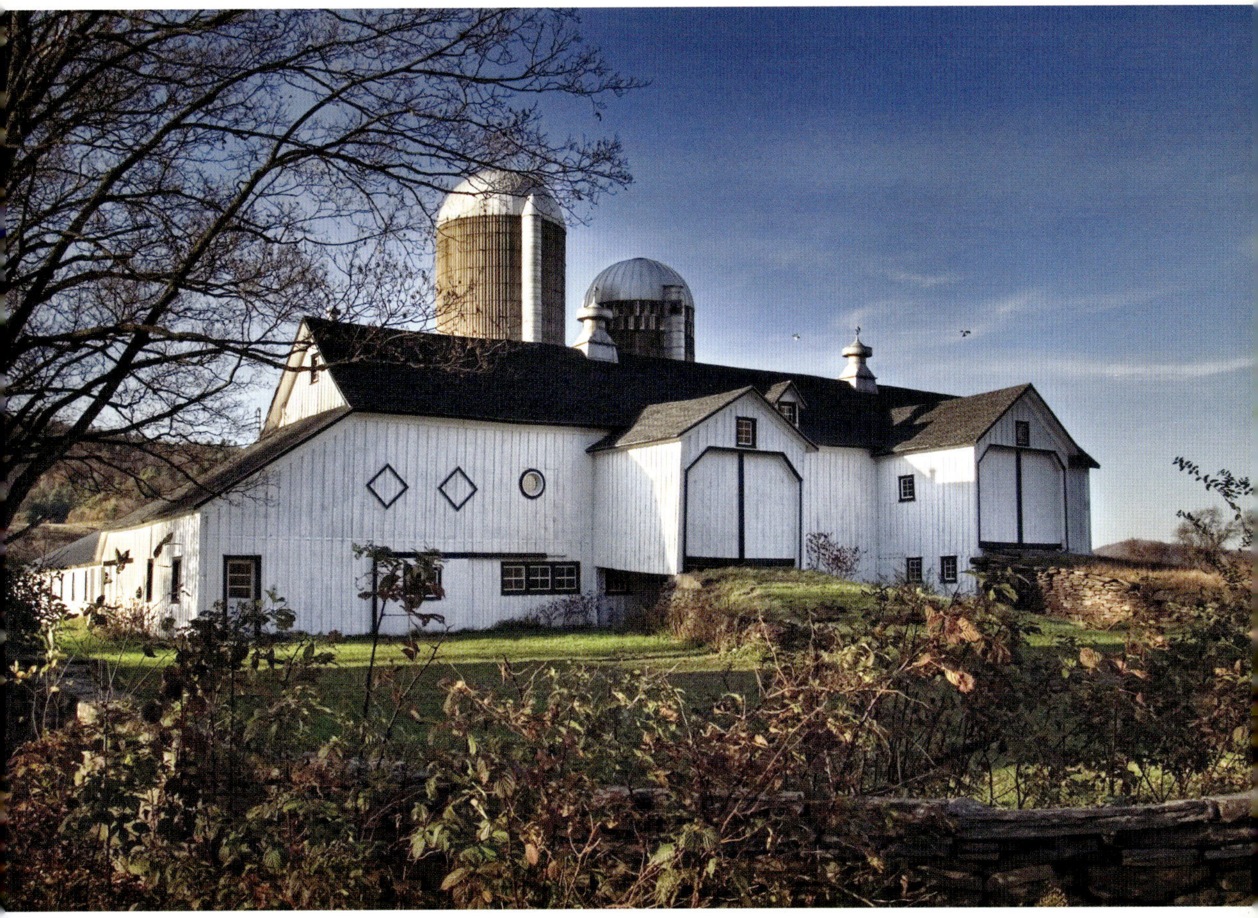

**top**—High-bridgeway barn. New York.

**bottom**—Double high-bridgeway barn. Catskills, New York

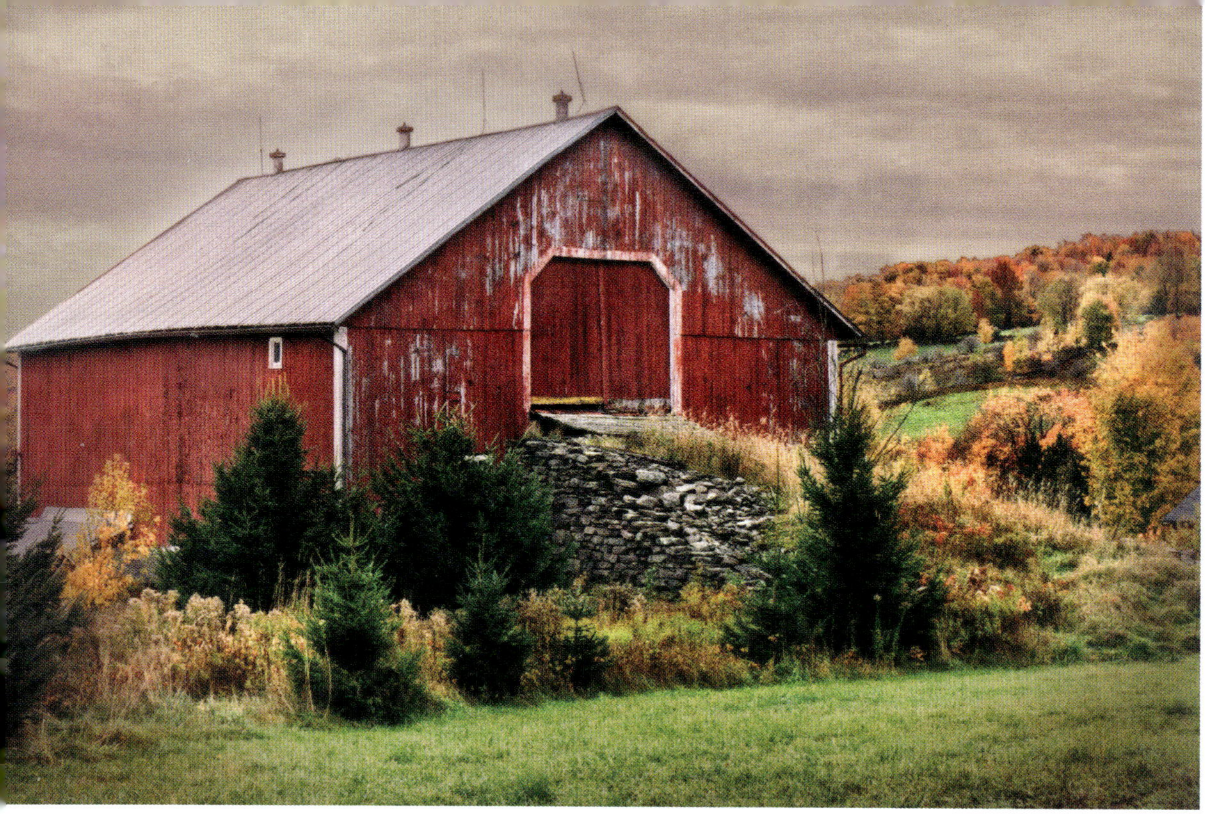

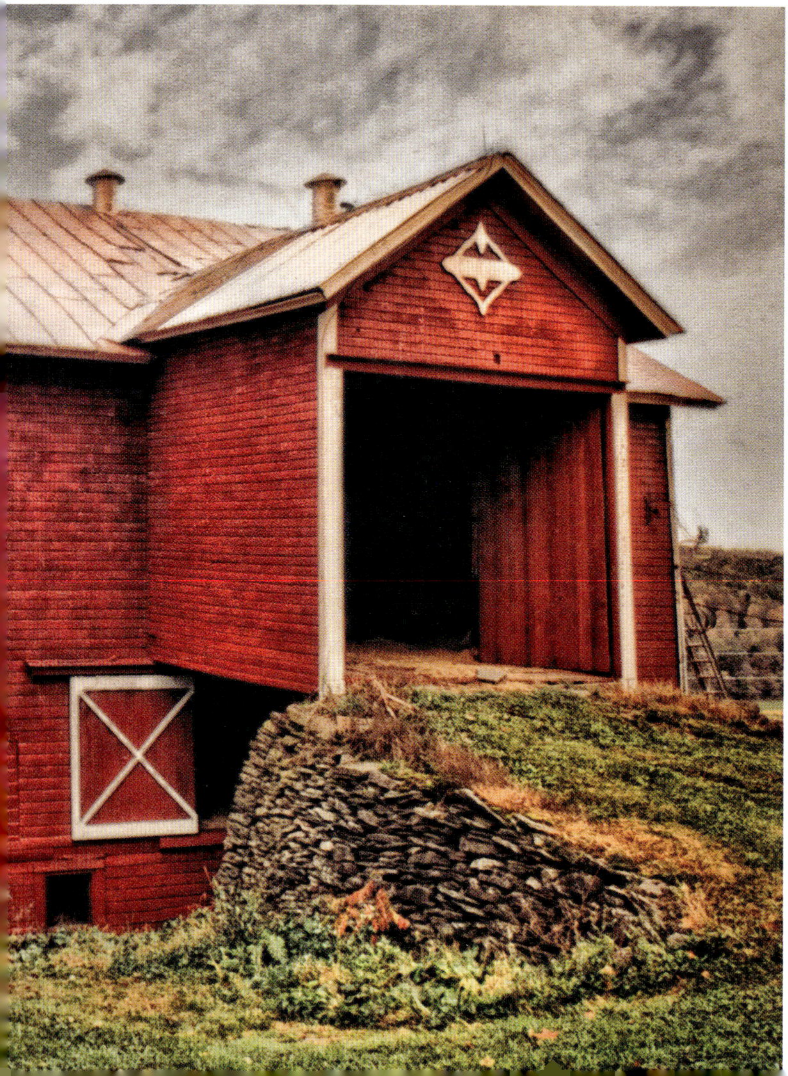

**above and left**—Gable entry, high-bridge way barn. Captured with a Mamiya RB67 (medium-format film camera) in 1985.

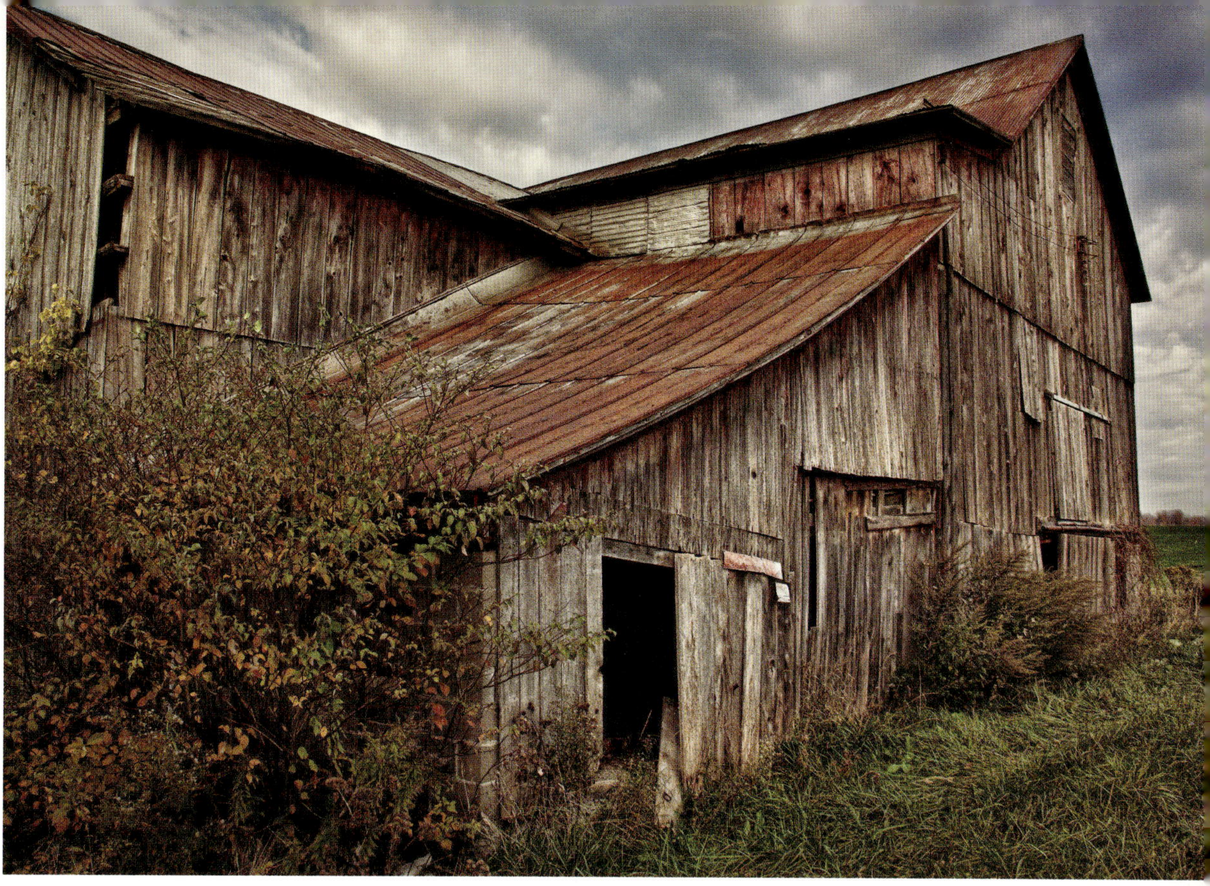

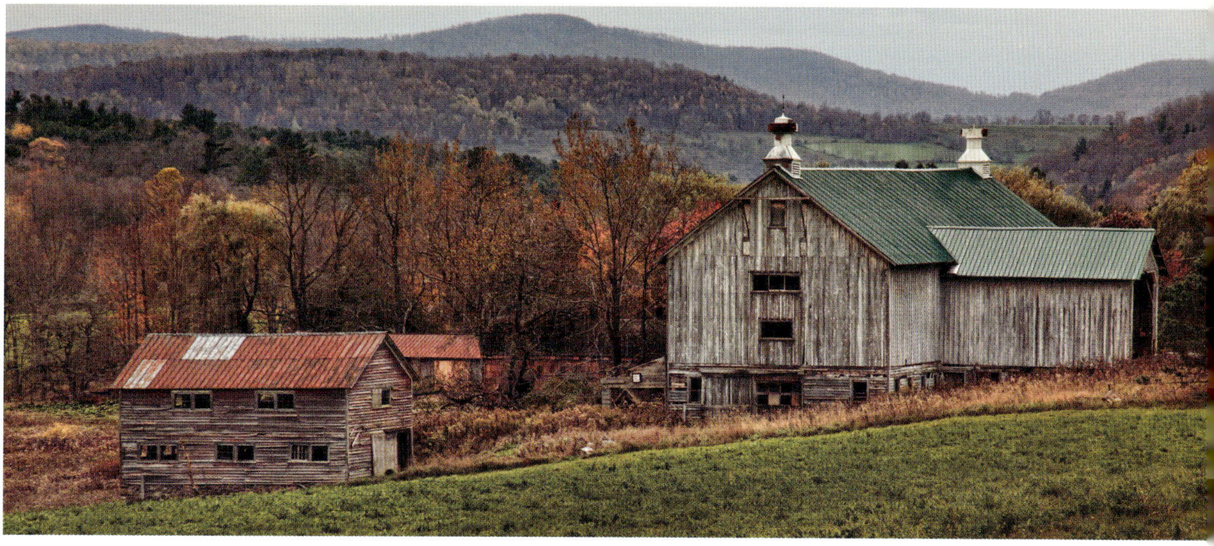

**top and bottom**—Bank barns. New York.

# Round Barns

Only a handful of round barns exist today. Most of these barns were built between 1890 and 1920, primarily in the mid-west and northeast.

Round barns were typically made of wood, though some were constructed with brick. A round barn was advantageous to farmers, as it had a greater volume-to-surface ratio than a rectangle barn and cost less to construct than would rectangular barns.

The open floor in round barns gave farmers more space in which to work. Cattle were kept on the ground floor, while feed was typically kept in the center, making it easy to distribute.

**below**—Round barn. Indiana, 1985. Shot on film.

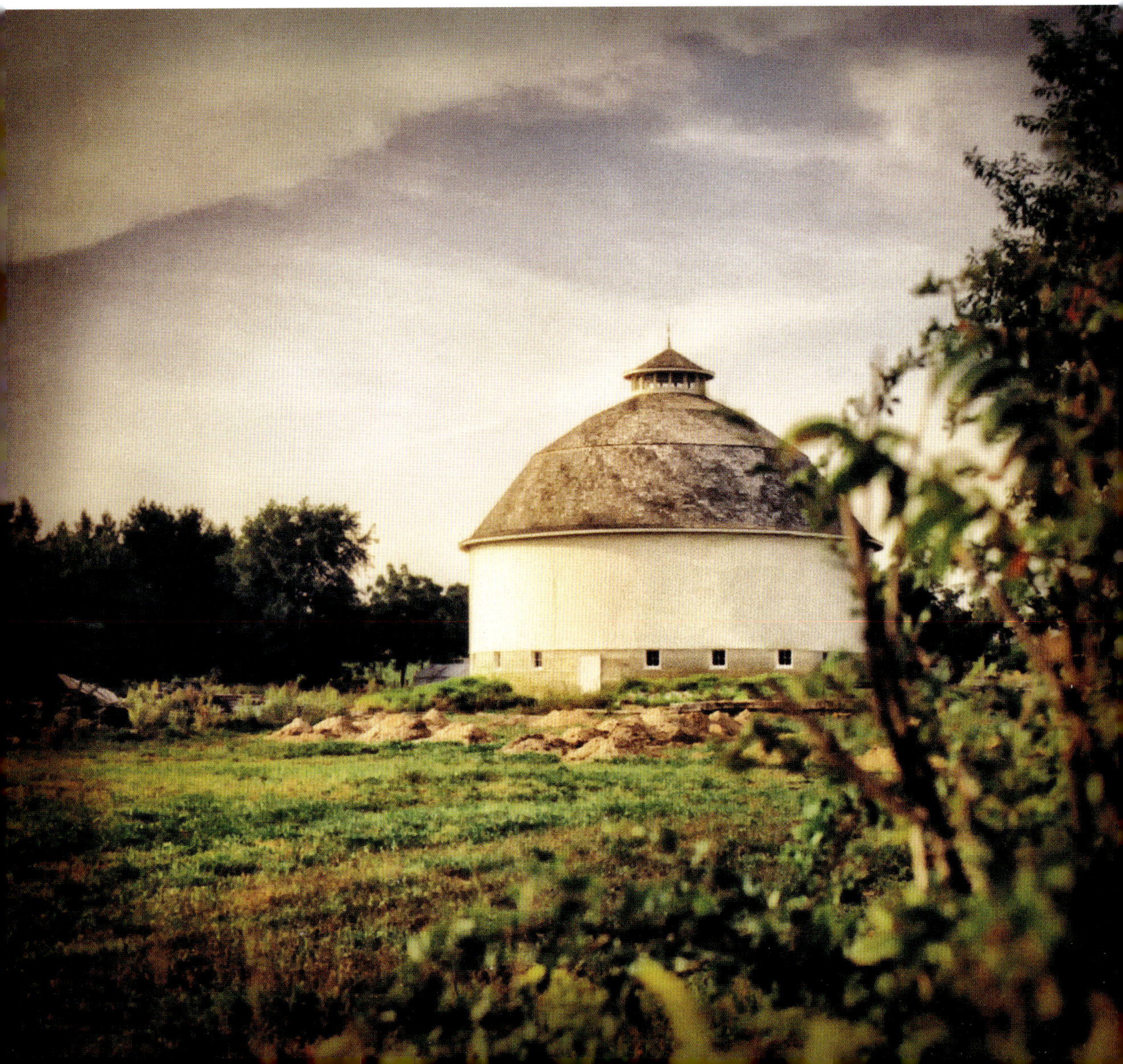

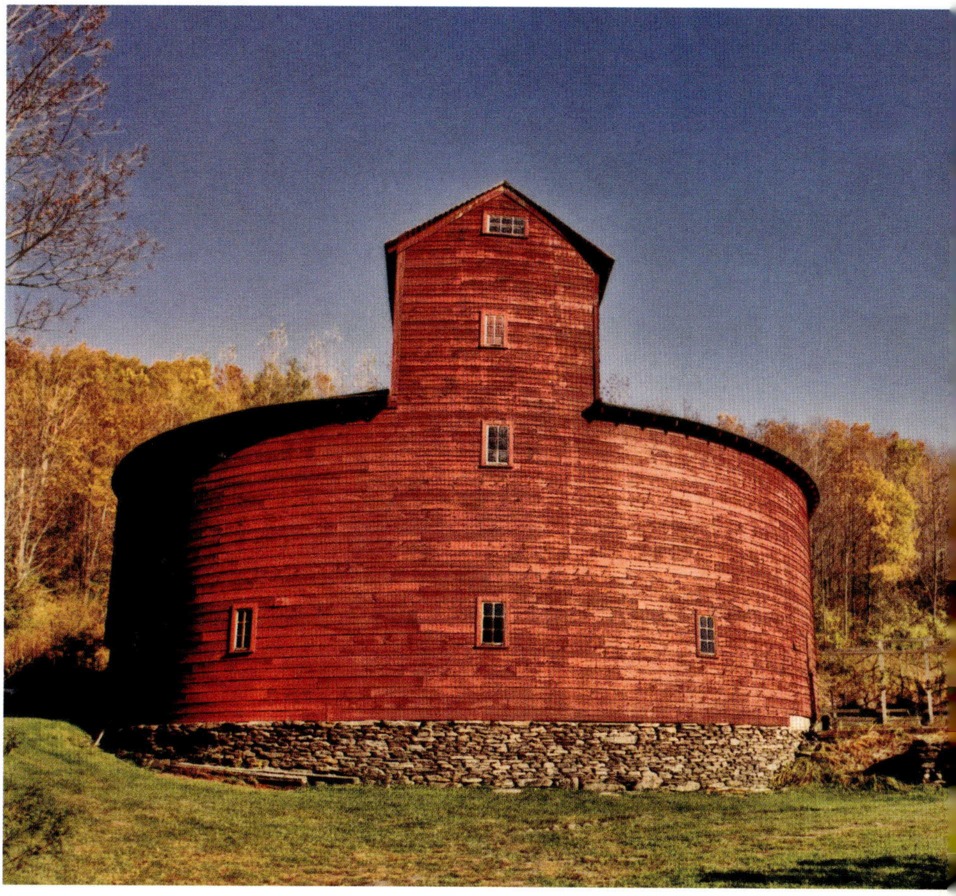

**right**—Catskill Mountain round barn, circa 1893. Shot on film.

**below**—Historic round barn. Gettysburg, Pennsylvania, circa 1914.

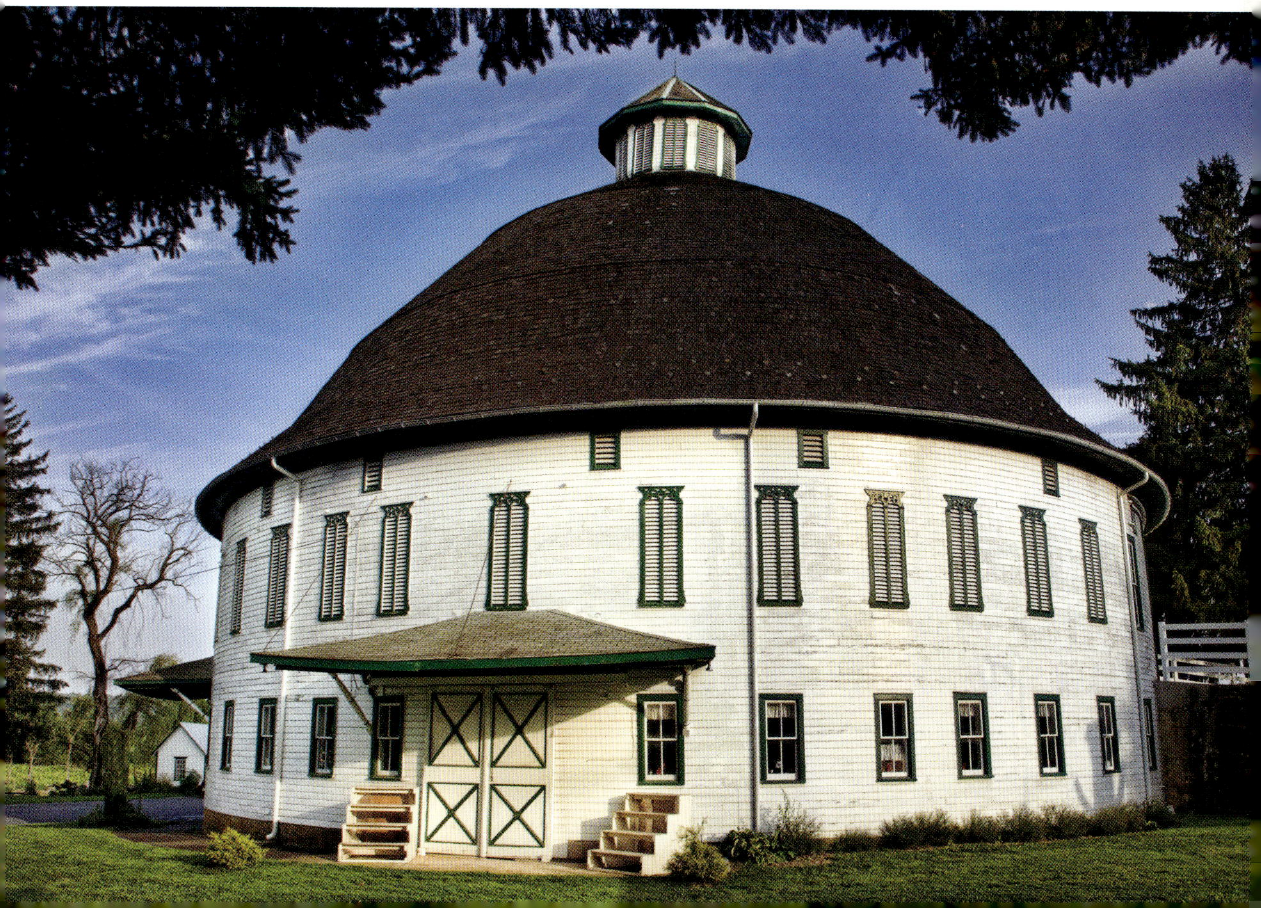

# Tobacco Barns

As you drive the back roads of Amish Country in Lancaster, Pennsylvania, you can smell the tobacco in the air. While a common sight in Pennsylvania, tobacco barns and tobacco sheds can be found across the east and south.

Tobacco barns were built in numerous styles depending on the type of tobacco grown. They were usually made of oak, poplar, or other regional timber, and some date back almost four centuries.

## Drying and Curing

Tobacco barns had slits running vertically along the walls to circulate the air, as ventilation was important for the animals and grain. Inside the barn or shed, the tobacco was hung across racks to ventilate and dry.

Tobacco is cured at specific temperatures. Cigar companies often have their own curing recipes. Traditional black Kentucky tobacco barns were painted black to raise the heat inside, aiding the curing of the tobacco.

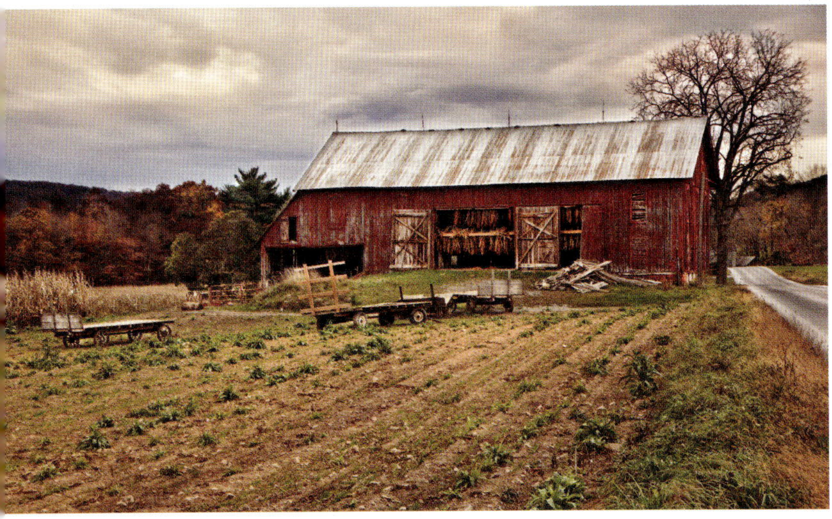

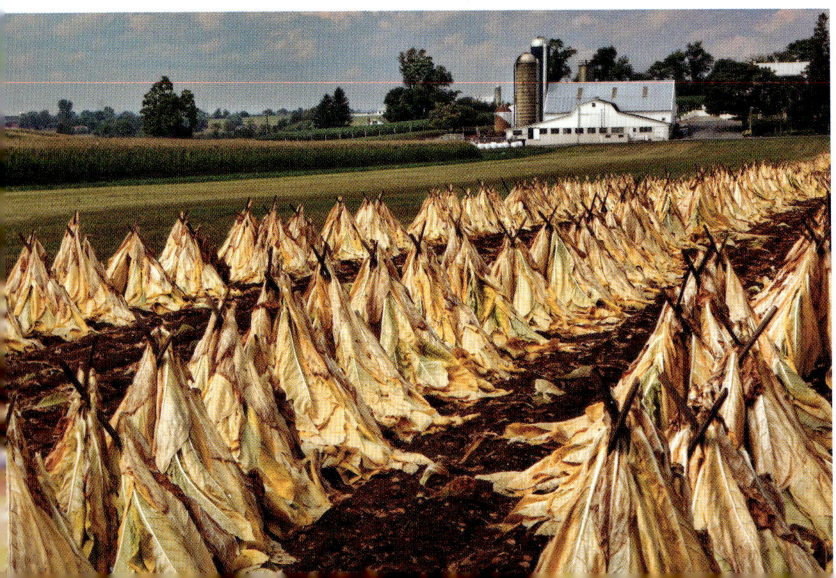

**top**—Pennsylvania.

**bottom and following page**—Tobacco farms. Lancaster, Pennsylvania.

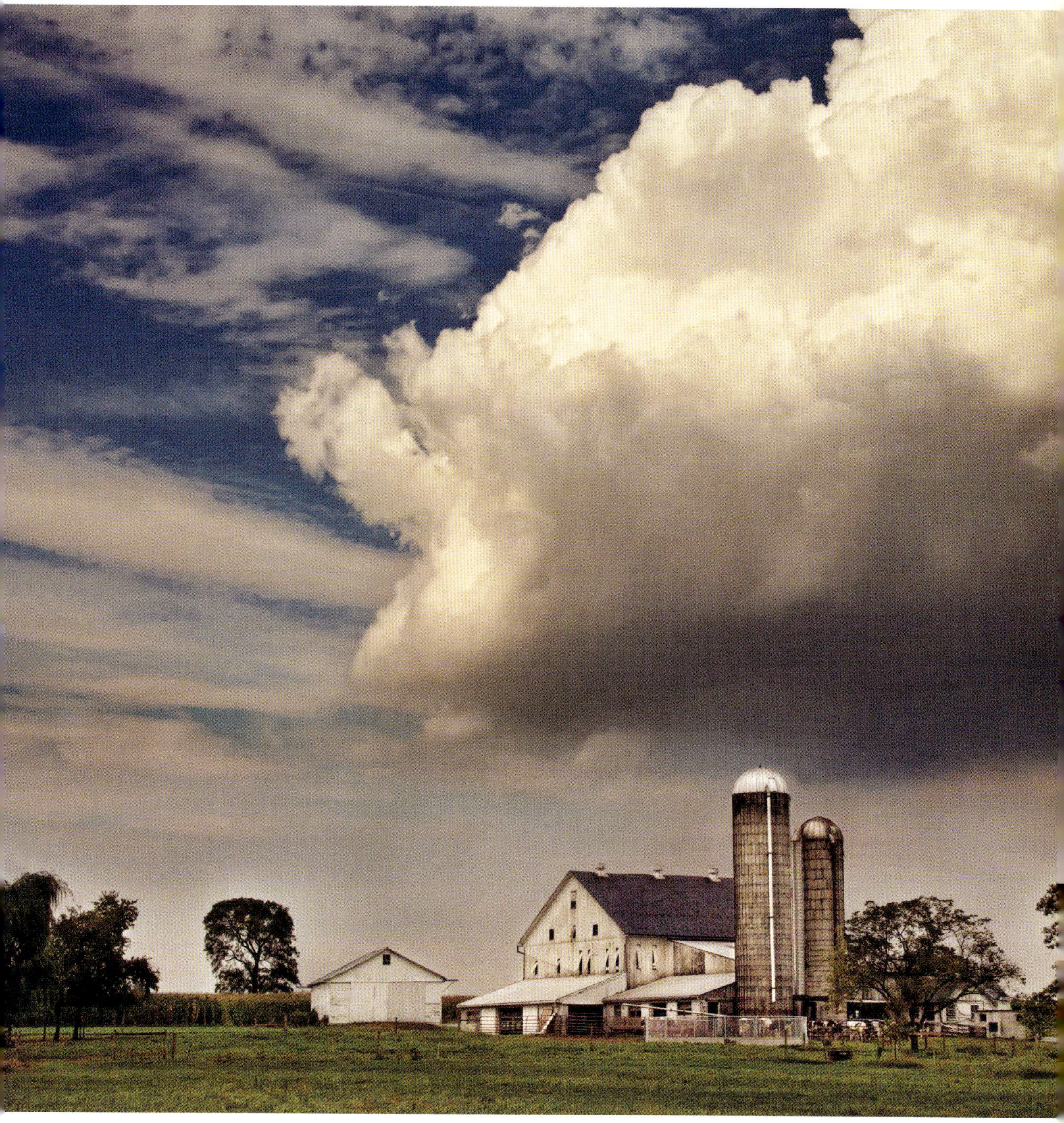

"Light makes photography. Embrace light, admire it, love it, but above all, know light. Know it for all you are worth, and you will know the key to photography."

—George Eastman

**This page spread**—Ventilated walls. Lancaster, Pennsylvania.

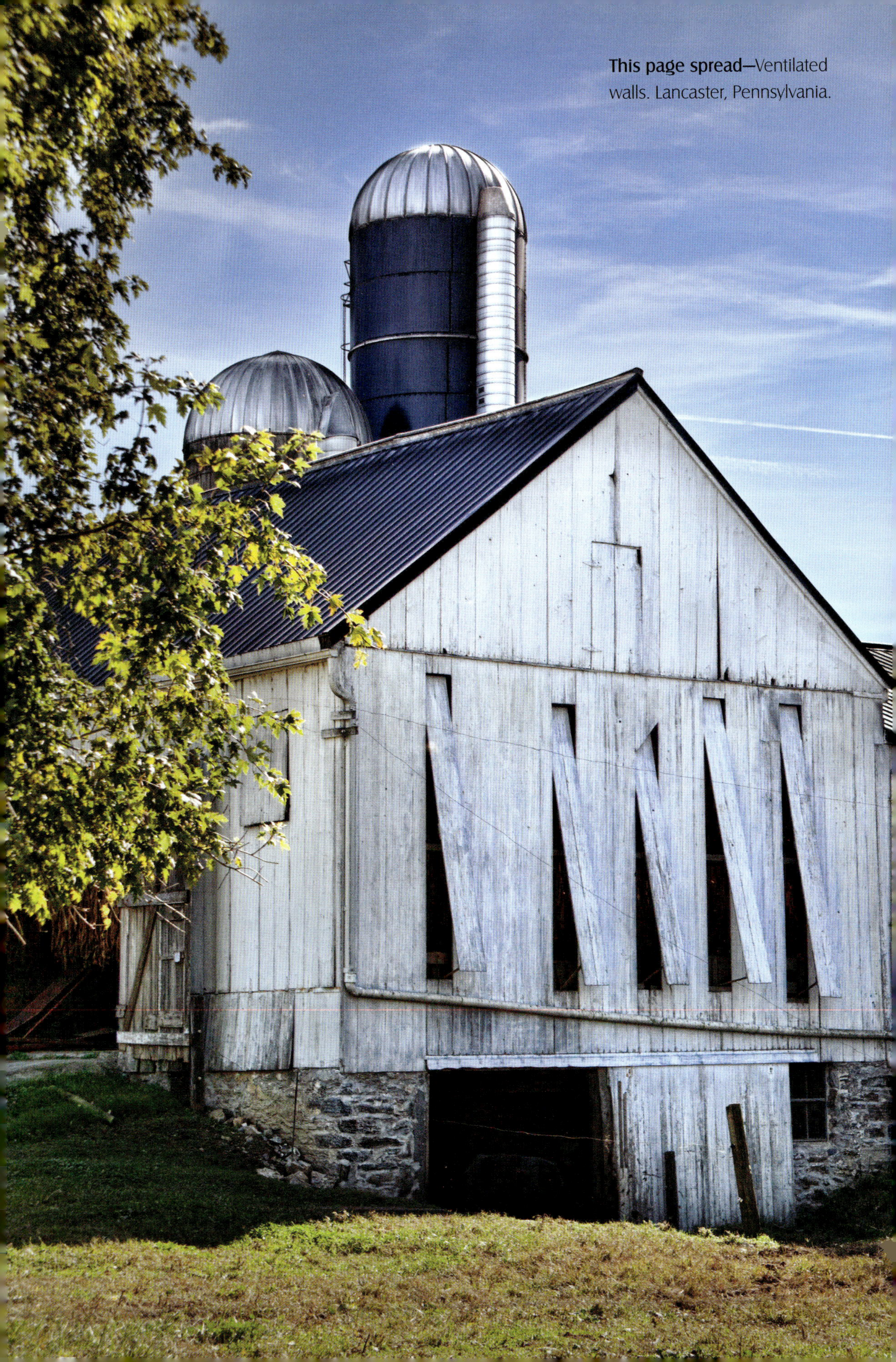

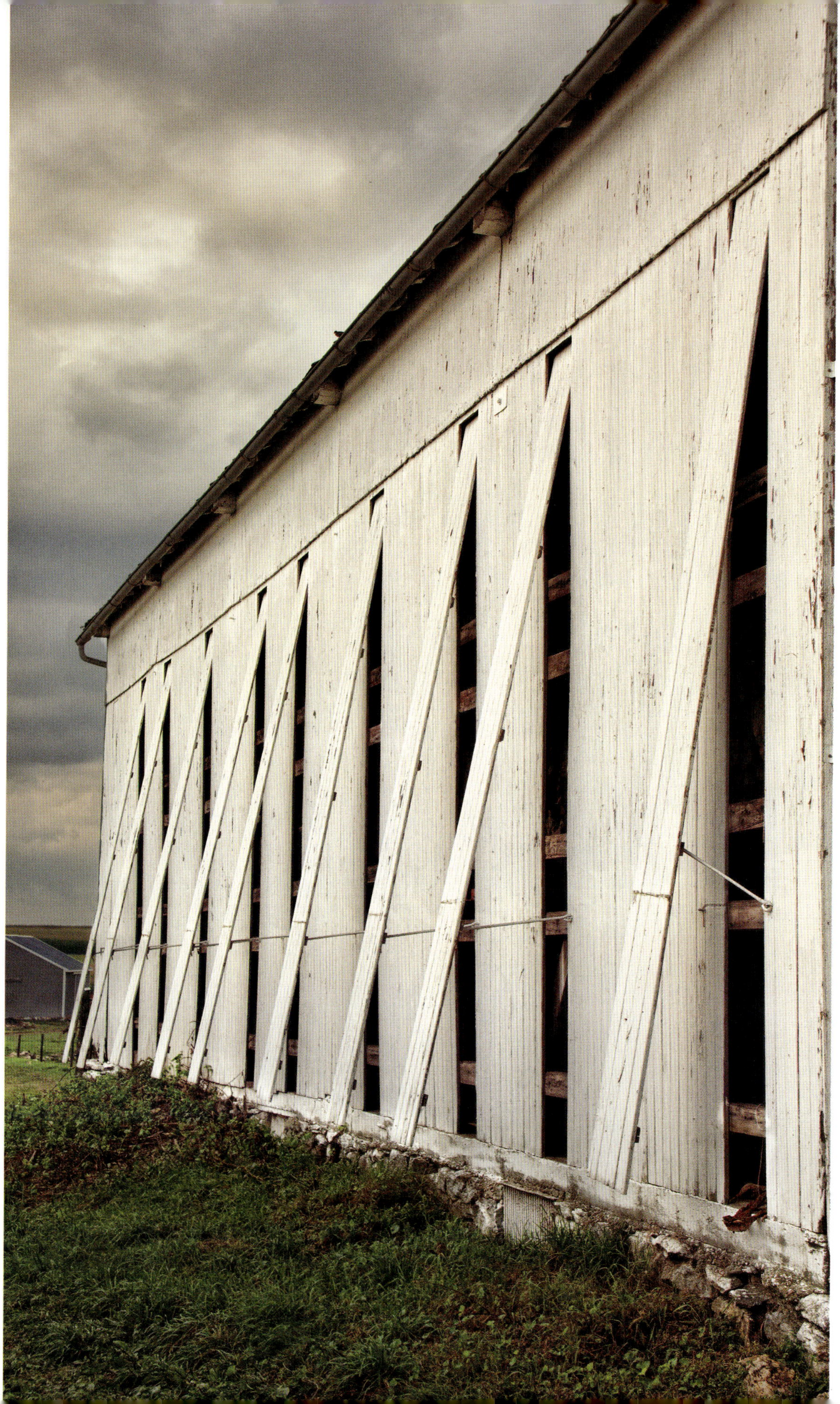

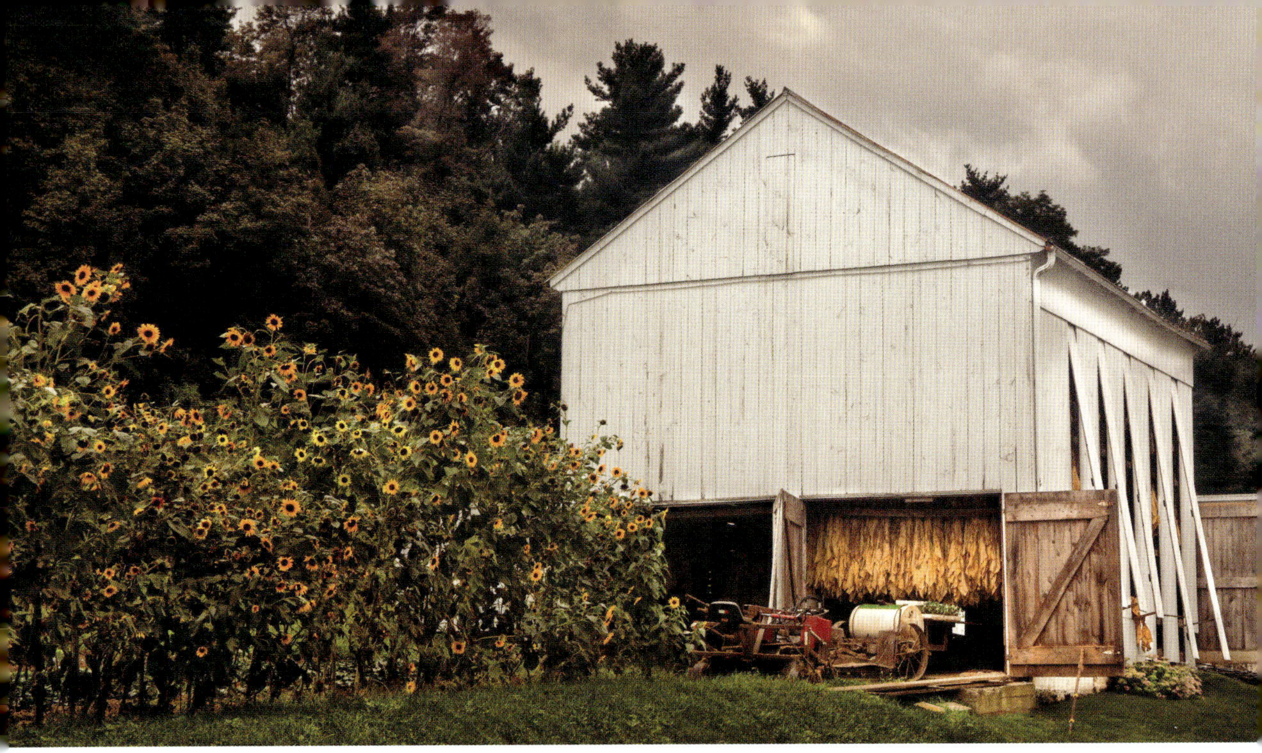

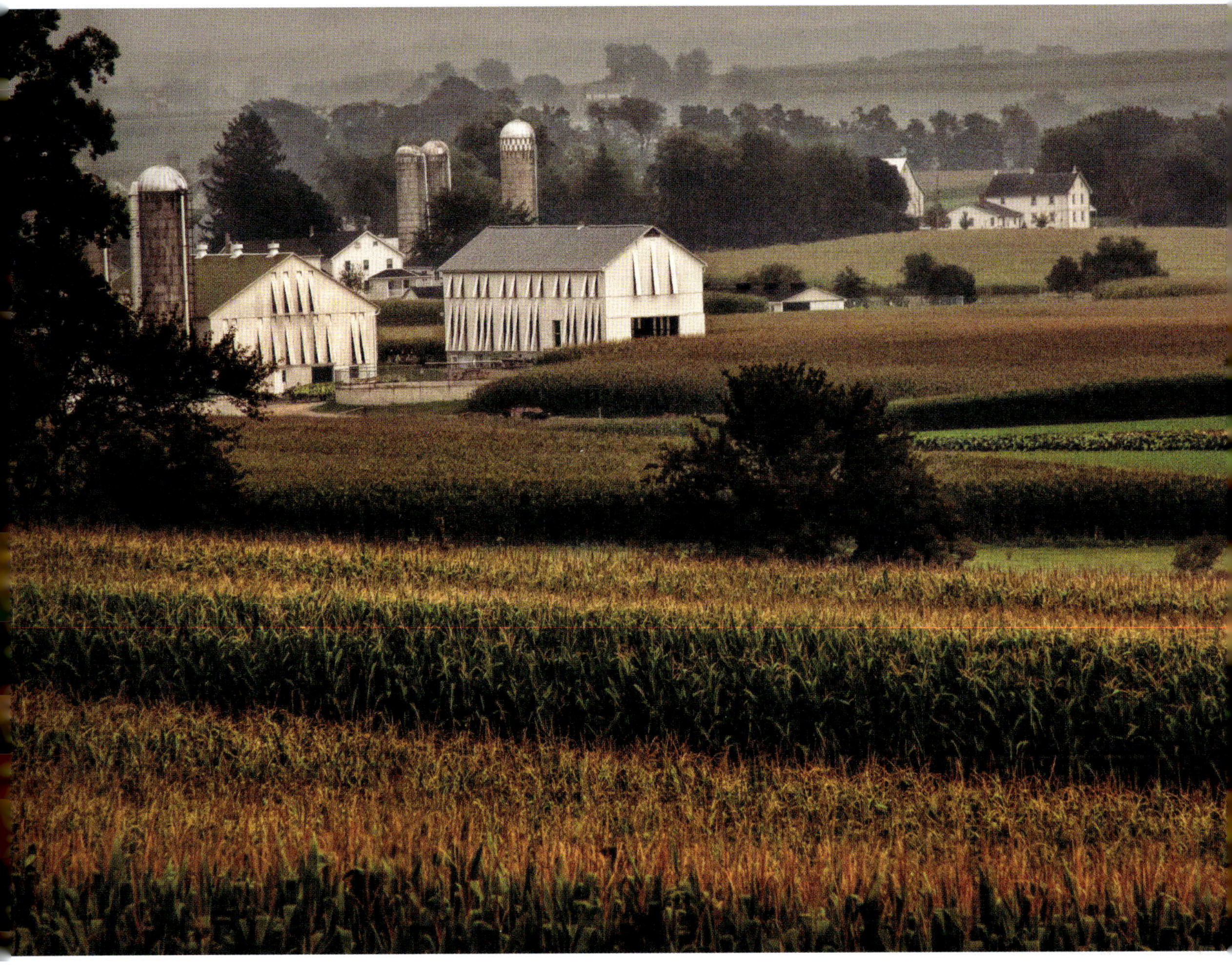

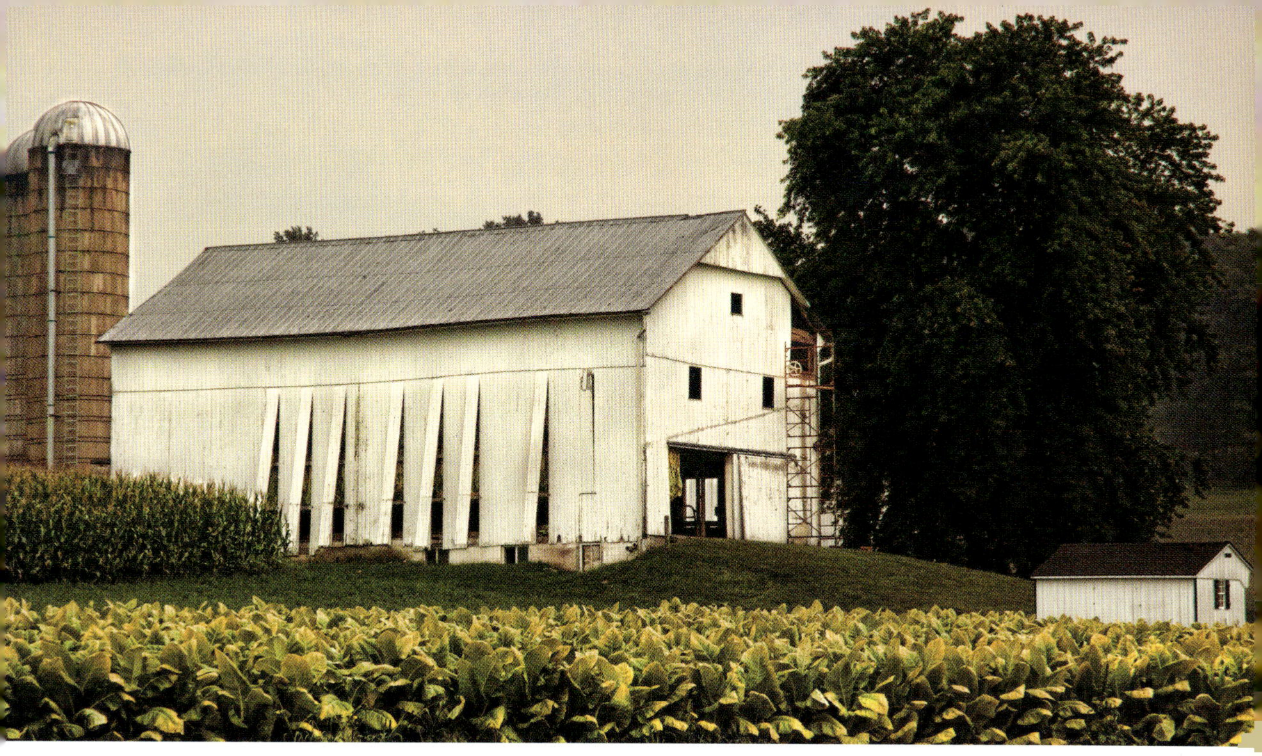

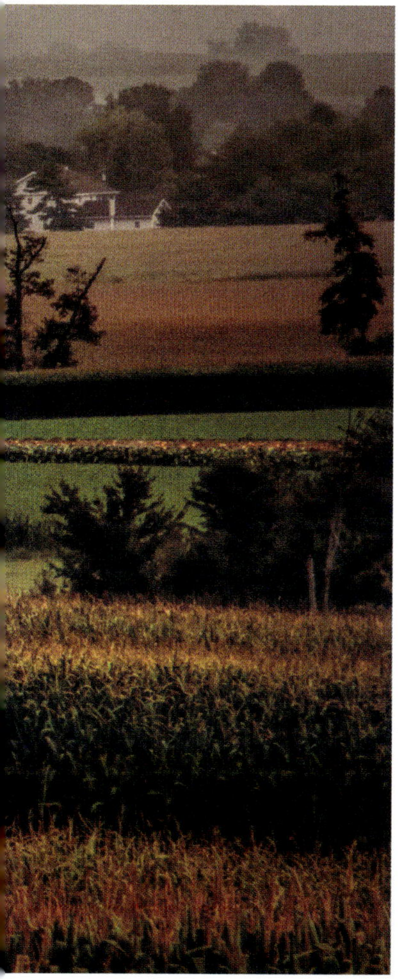

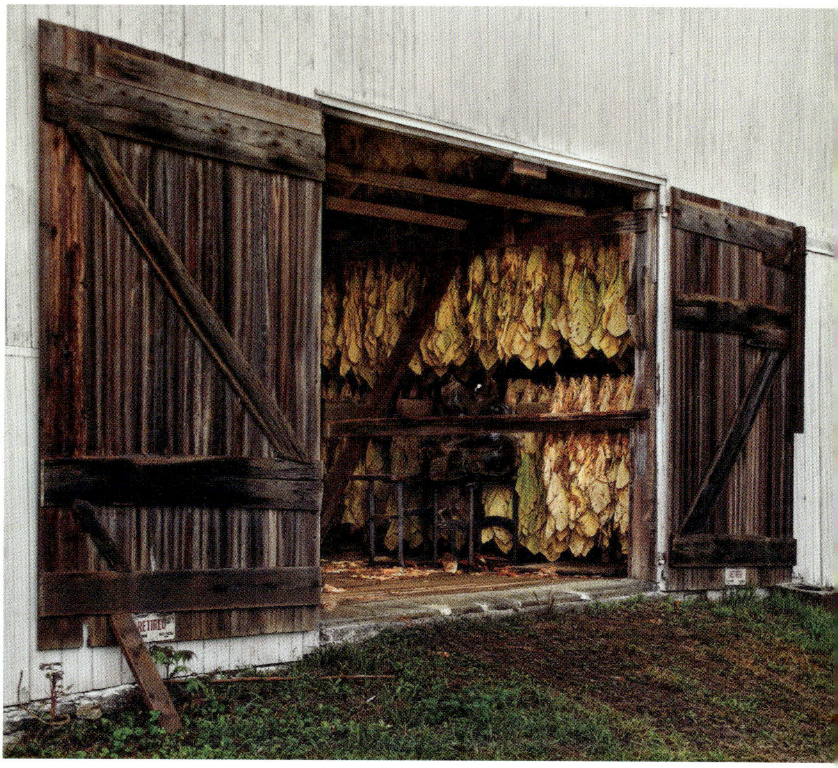

this page spread—Tobacco barns in Pennsylvania.

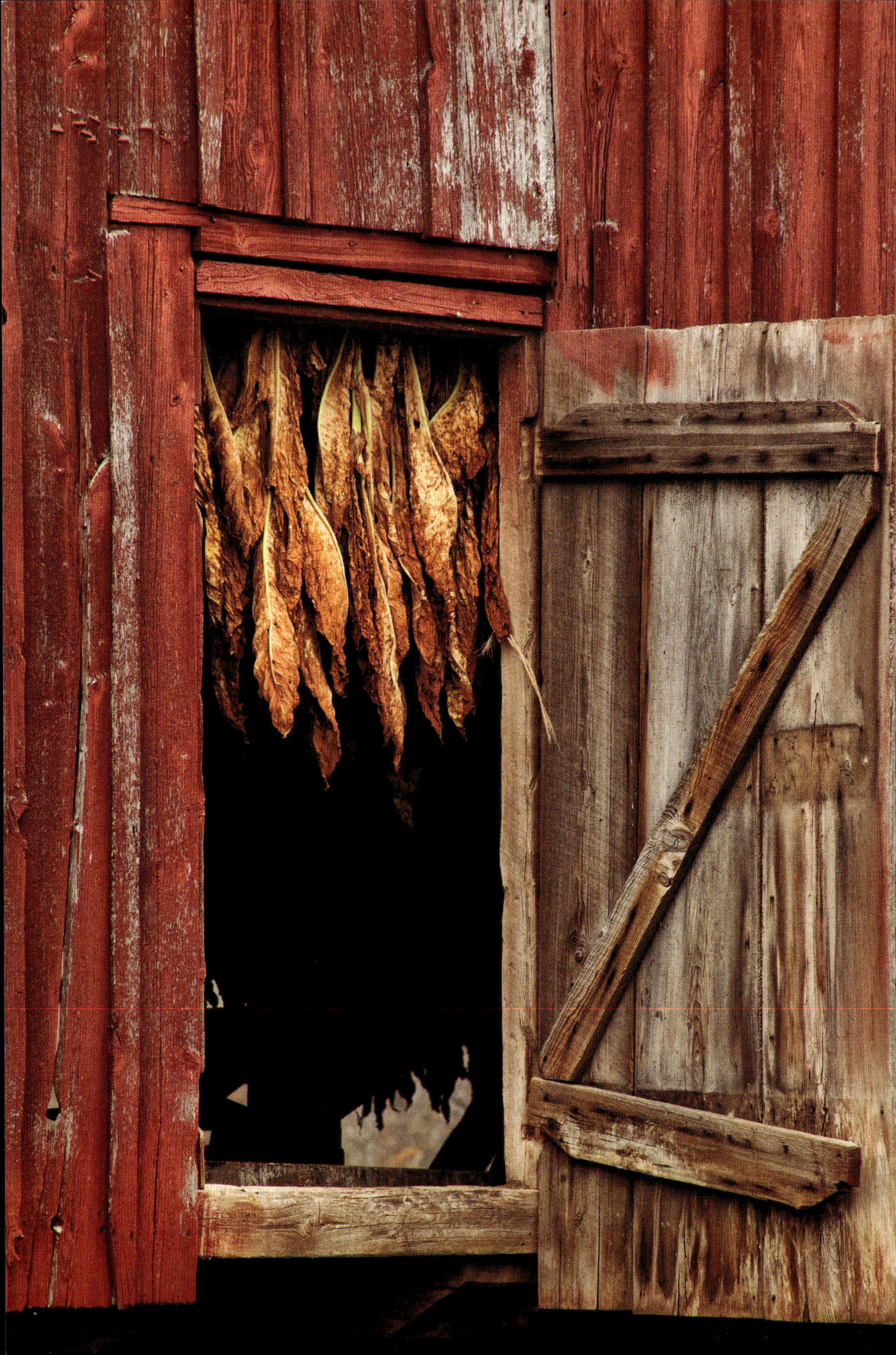

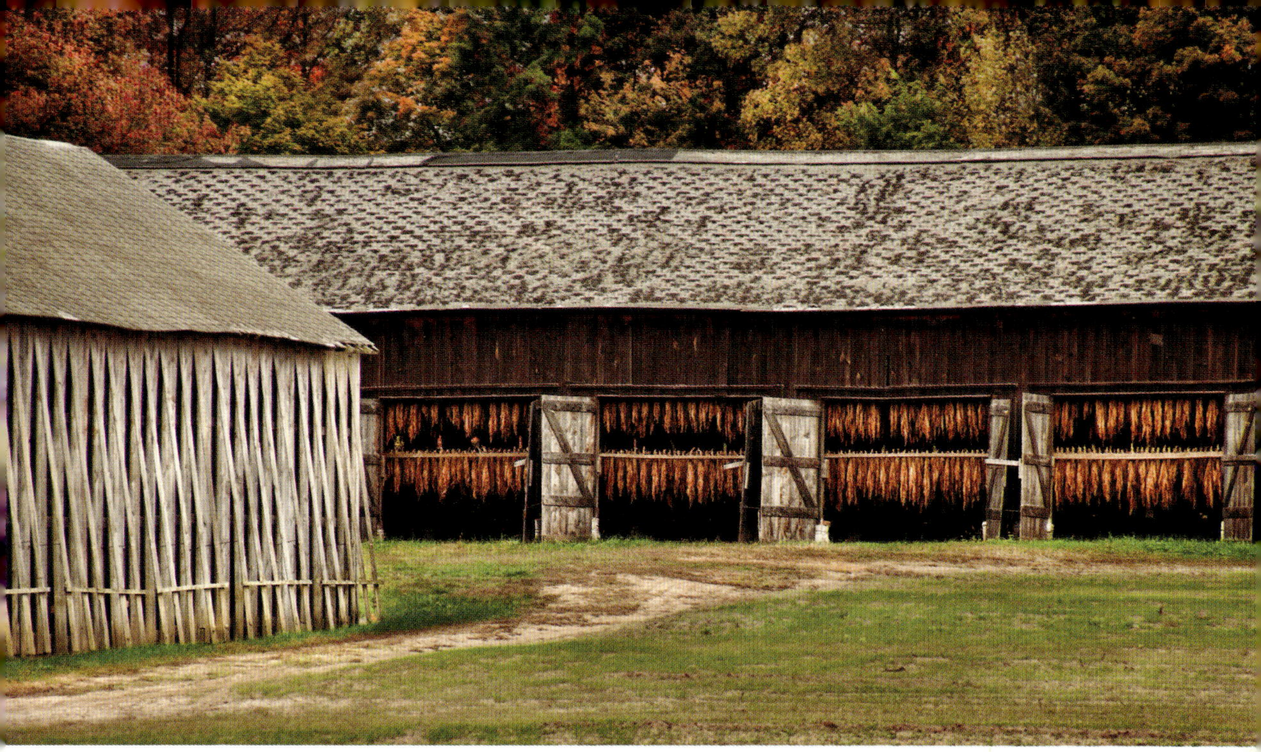

**previous page—** Tobacco drying in an Amish barn in Pennsylvania.

**above—**Tobacco shed. Connecticut.

**right—**Tobacco hung to dry in Pennsylvania.

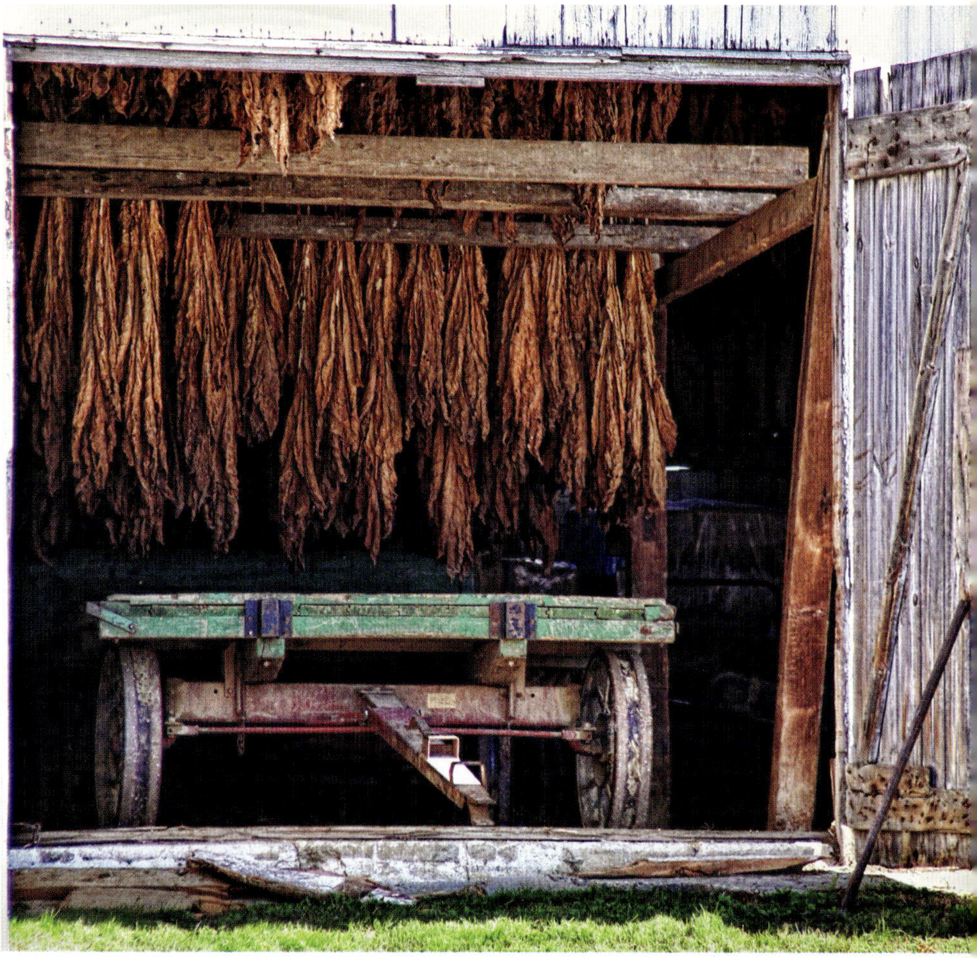

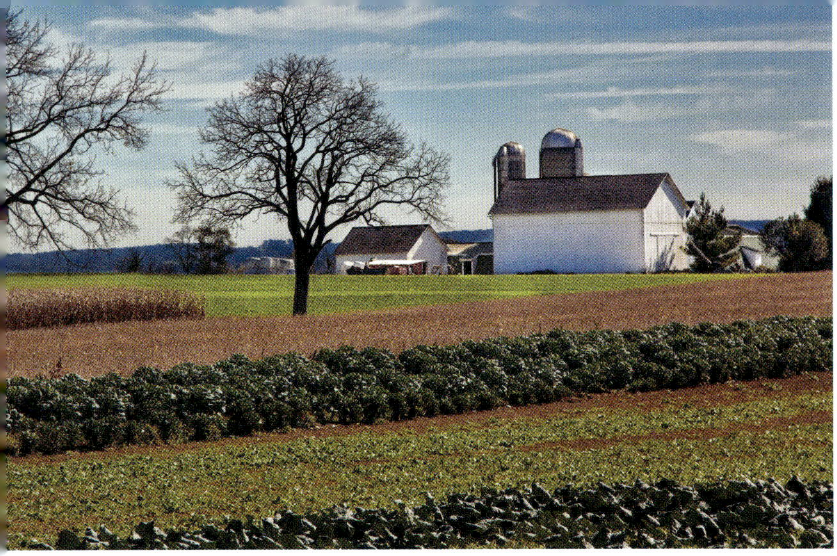

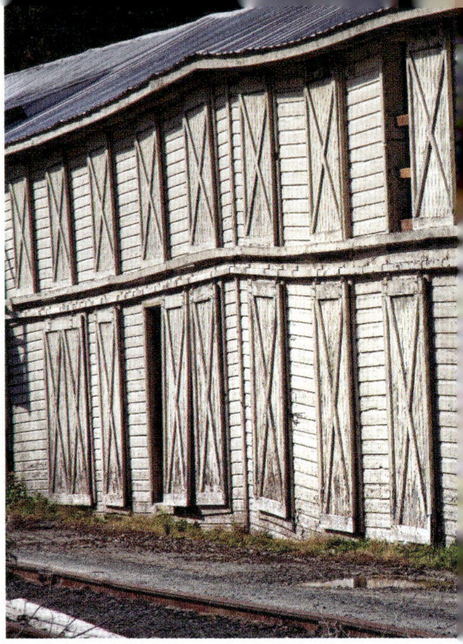

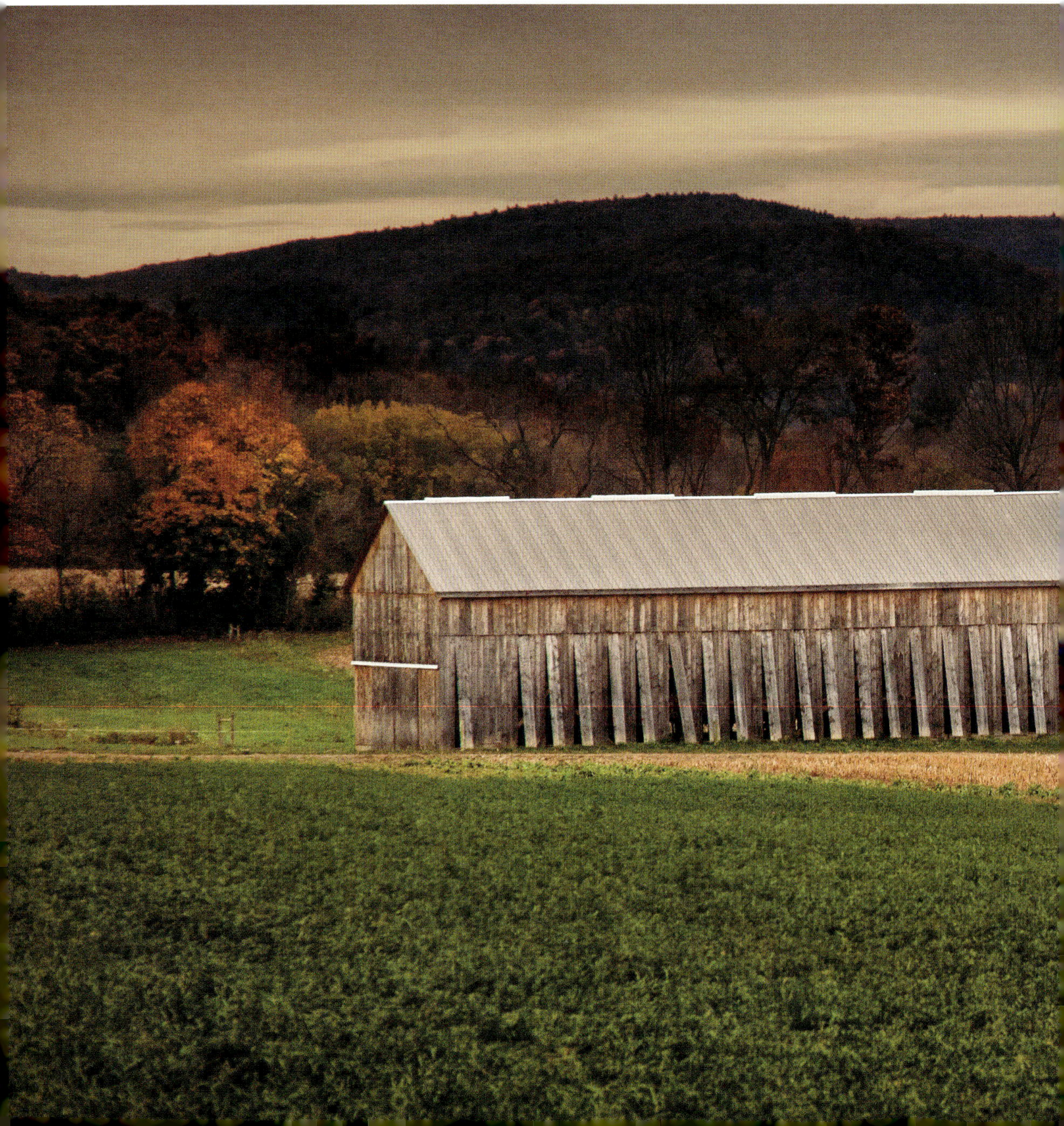

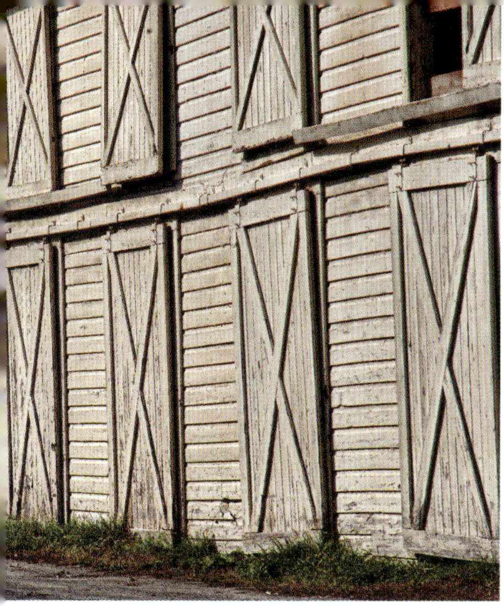

**previous page, top left**—Tobacco farm. Pennsylvania.

**left**—Tobacco shed located in Pennsylvania.

**below**—Tobacco barn, Pennsylvania.

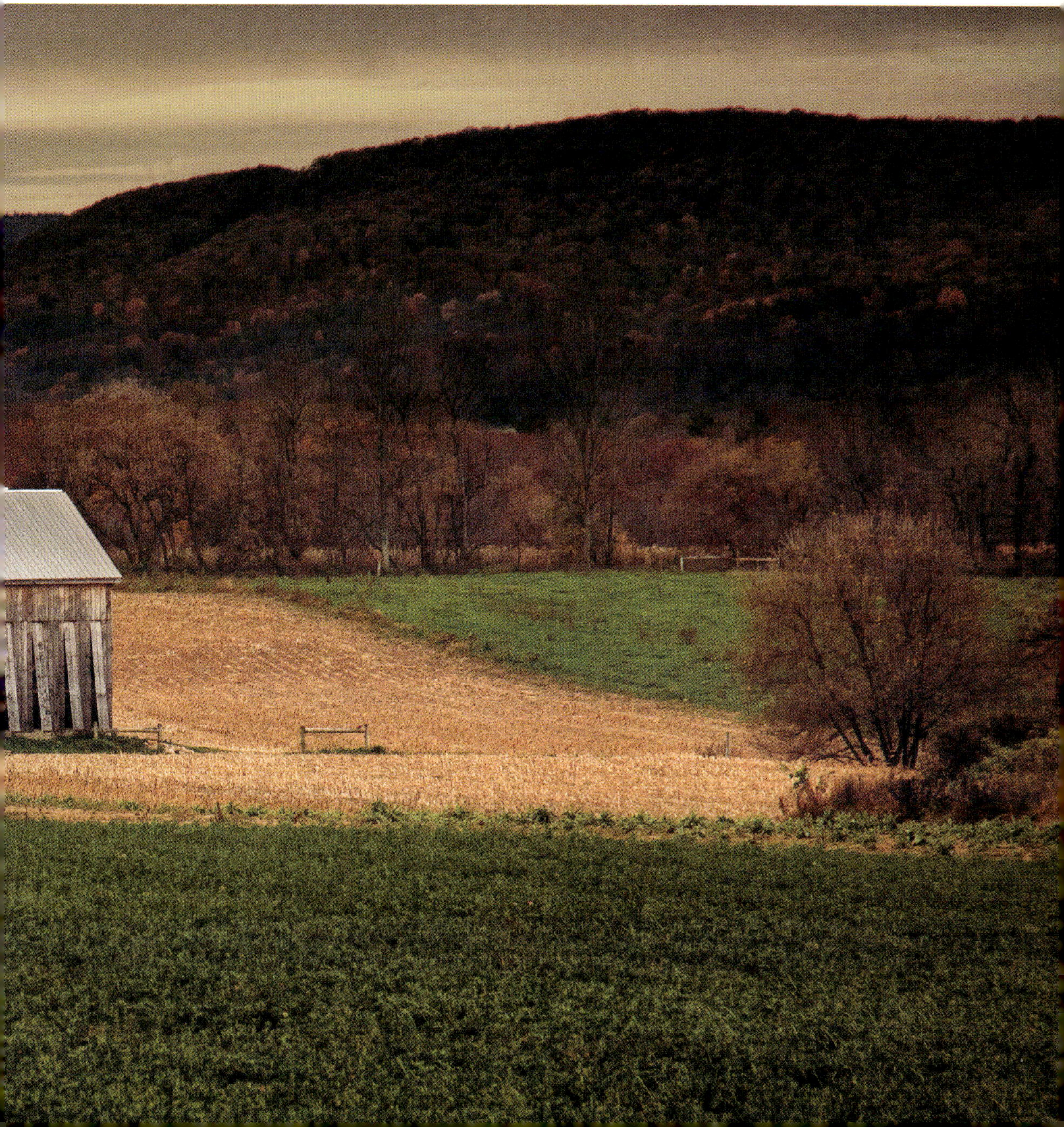

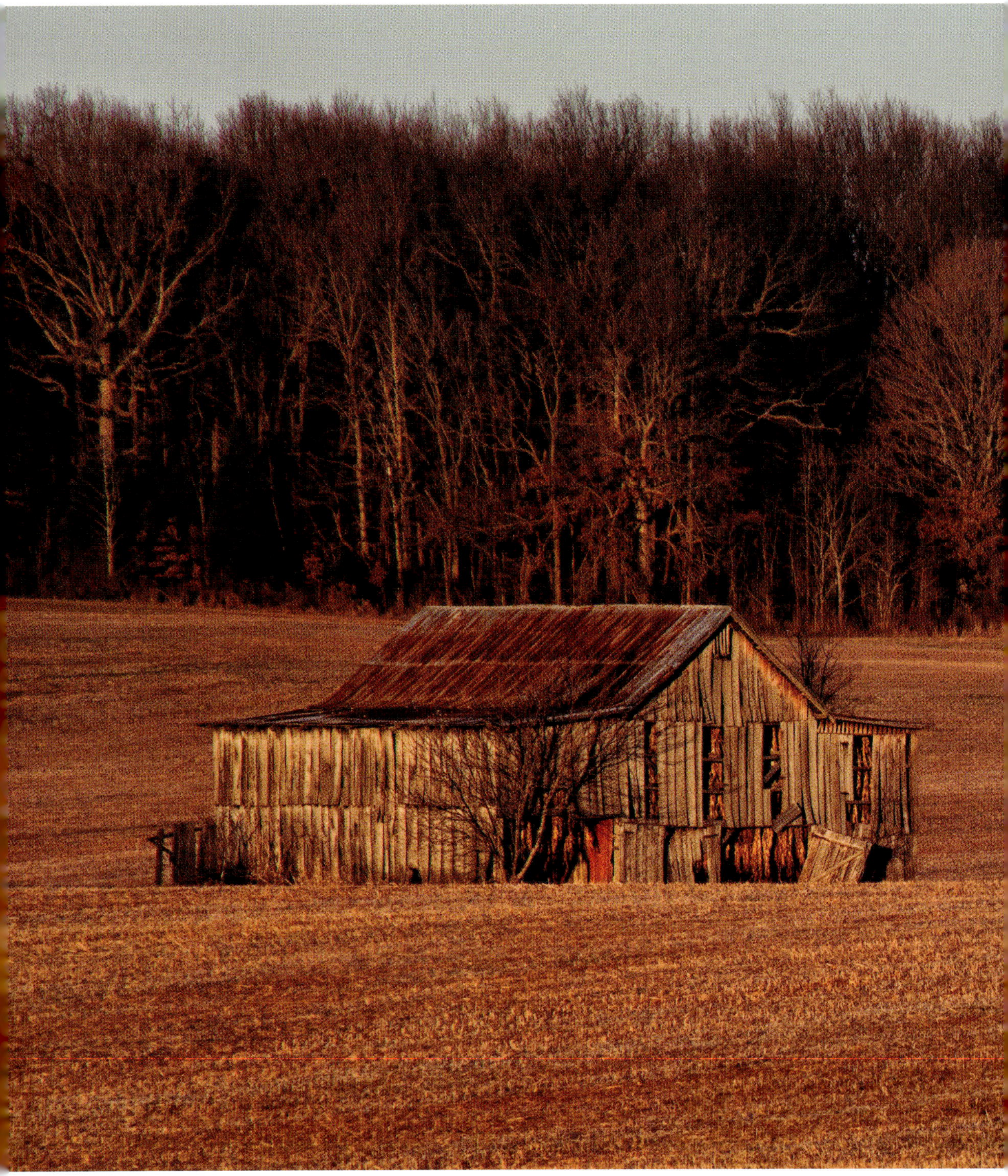

**above**—Tobacco barn. Kentucky.

# Dairy Barns

What's more iconic than a dairy farm, with cows peacefully grazing in the sun?

Not surprisingly, dairy farmers typically raised cattle to produce nutritious and tasty milk and other dairy products. In the early 1900s, there was a surge in dairy farming. Pasteurization made milk safer to drink, extended its shelf life, and allowed it to be transported greater distances.

## Just Add Chicken

Some dairy farms were "chickenized"—in other words, altered to accommodate chickens. Floors were added, and windows, too, to allow for natural light and increased ventilation.

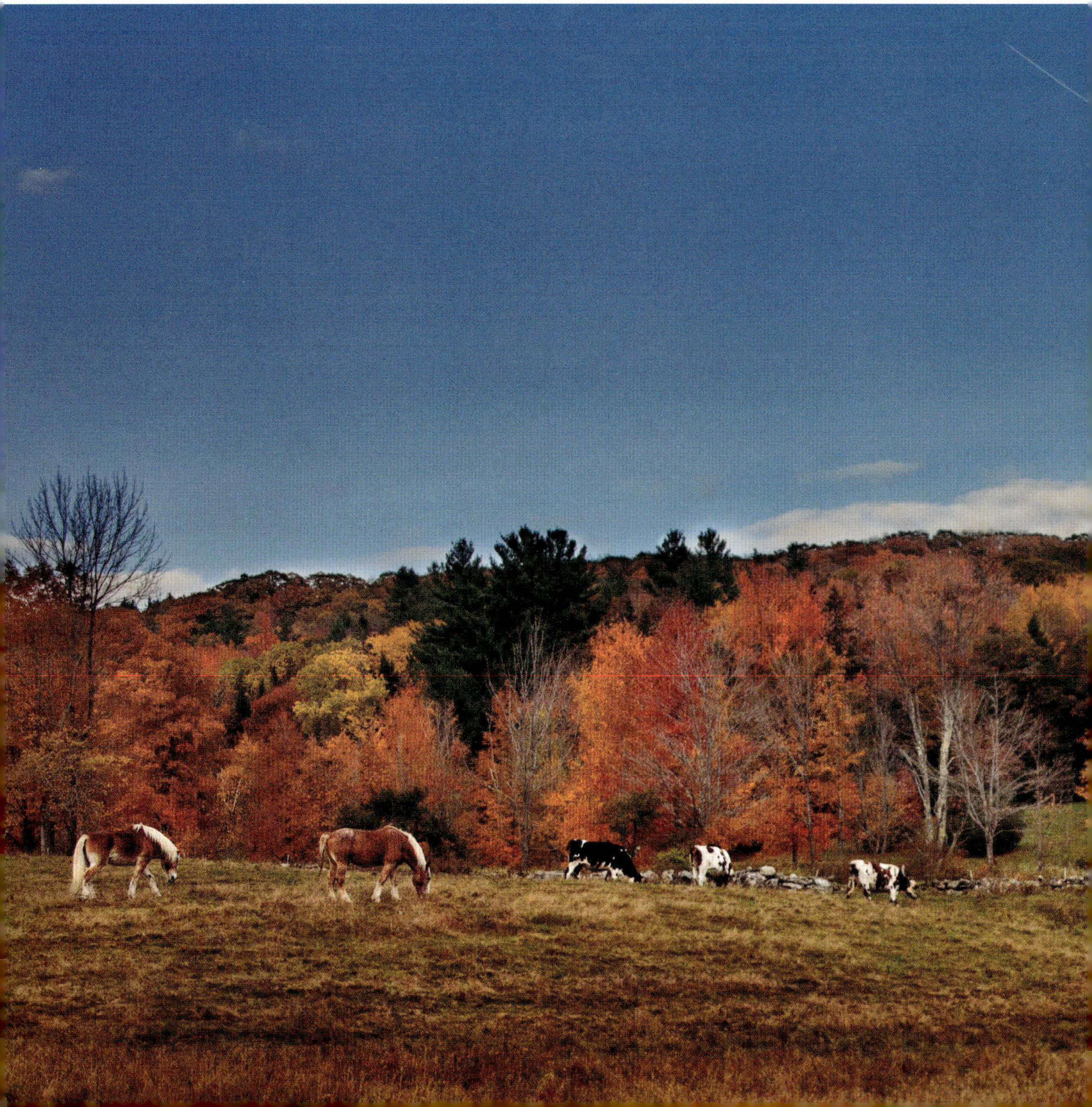

**right**—Photographed on film. New York.

**below**—A pair of horses commune with a group of cows on this picturesque dairy farm in Vermont.

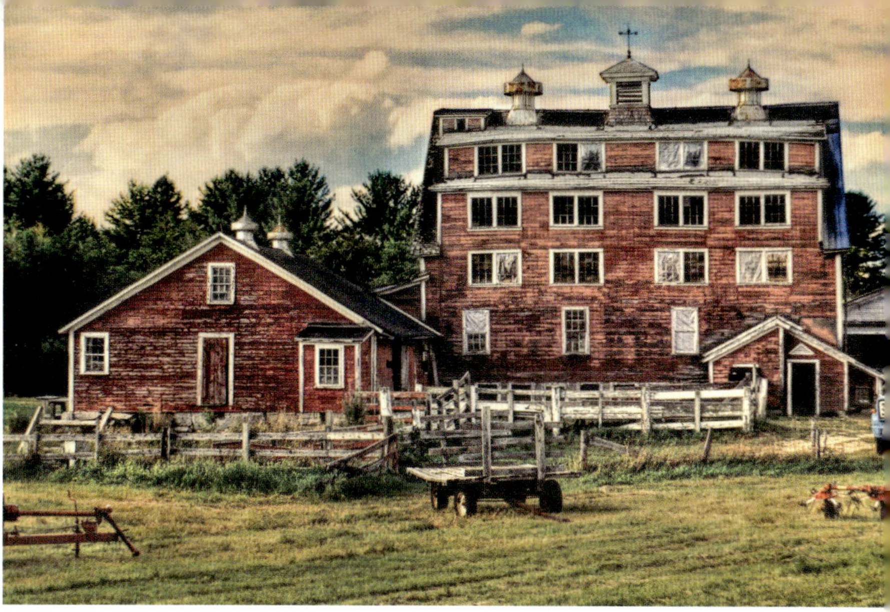

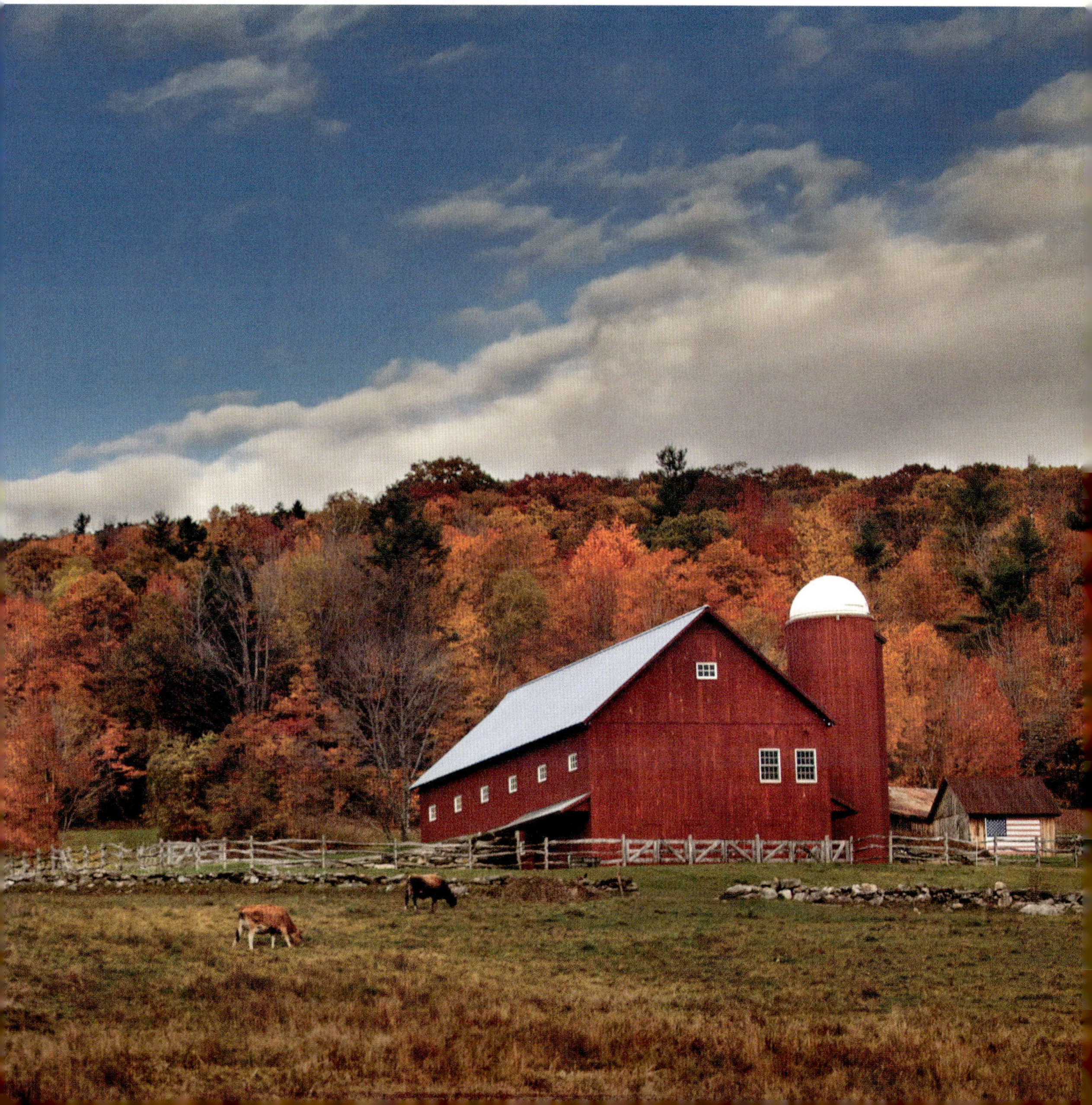

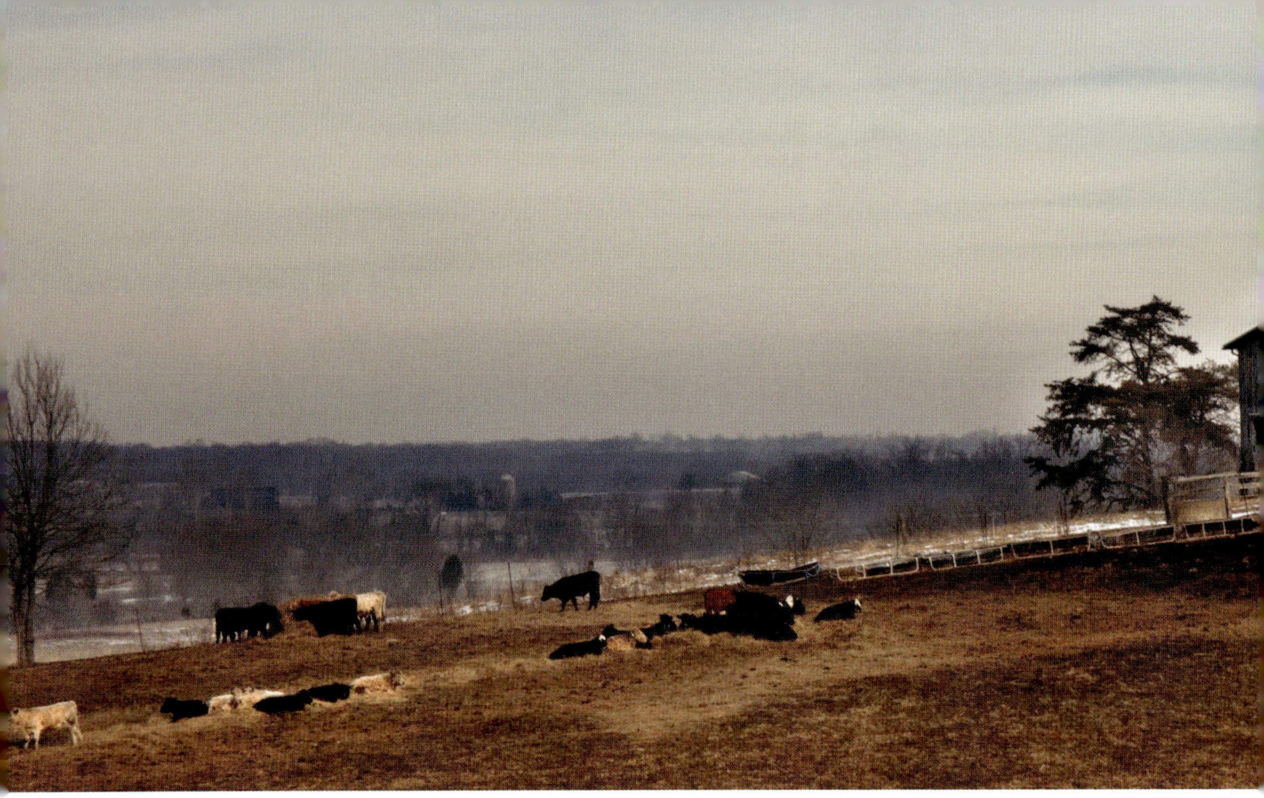

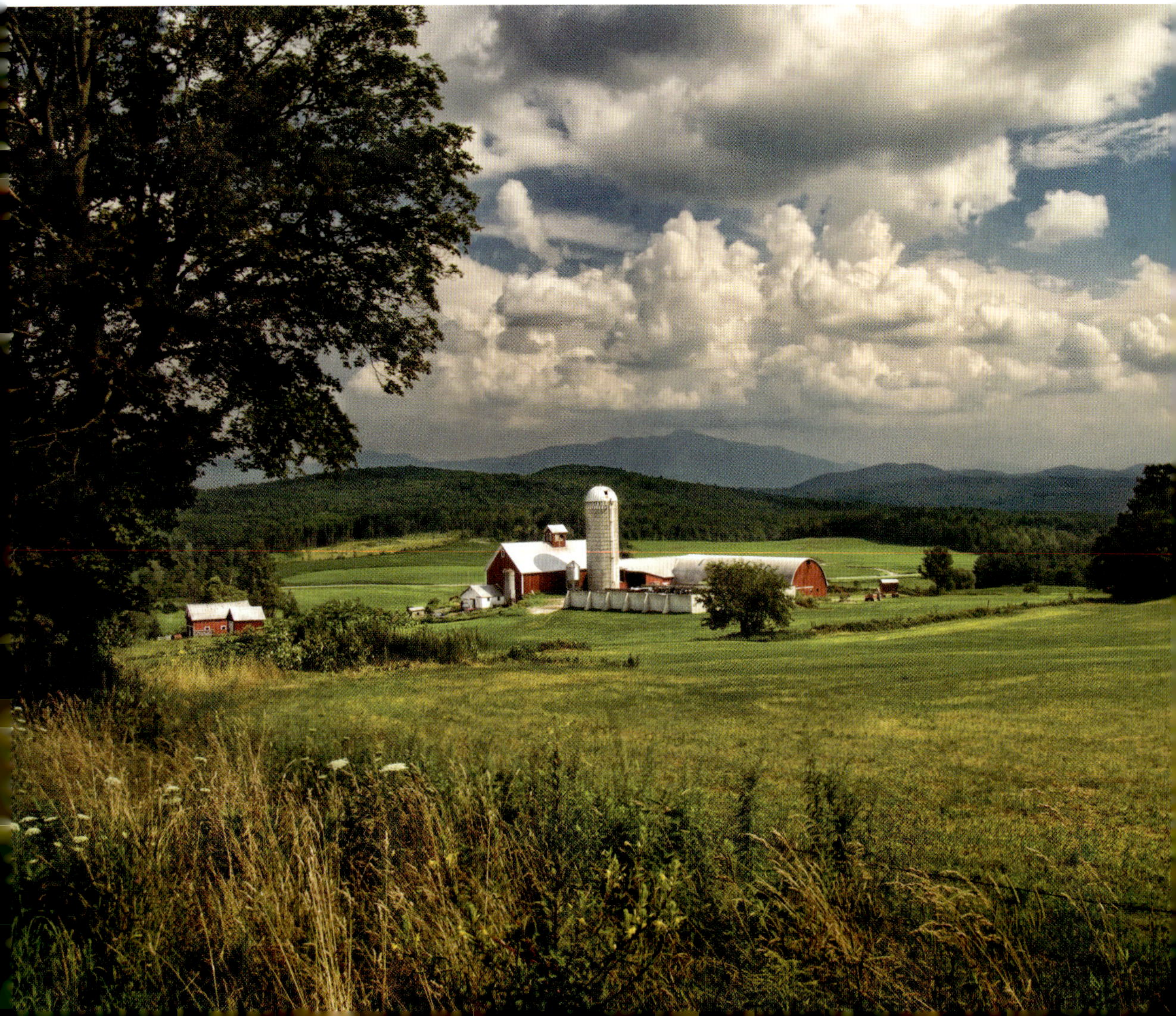

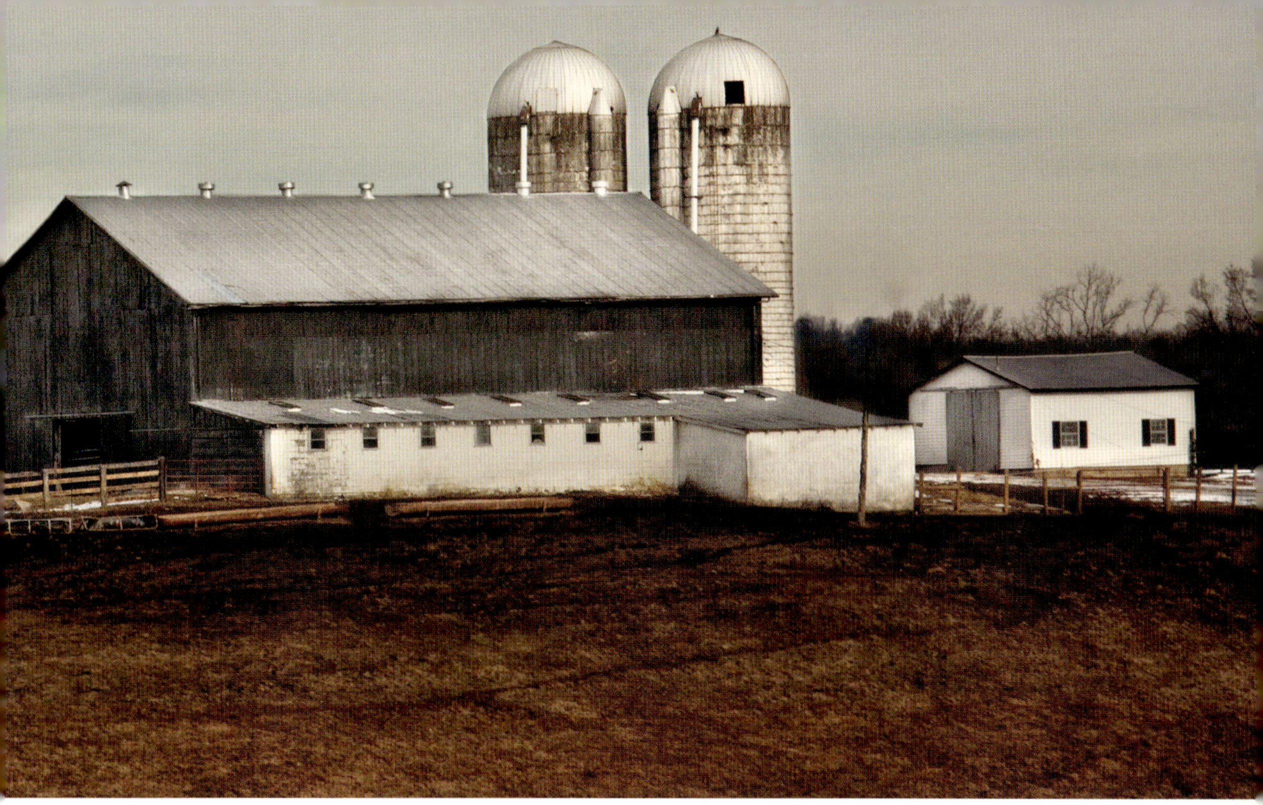

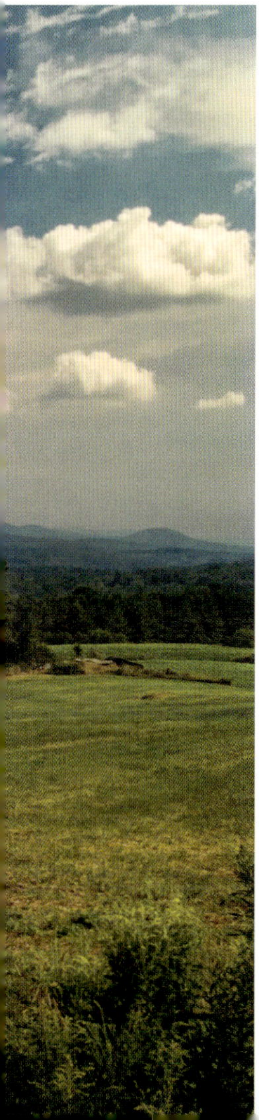

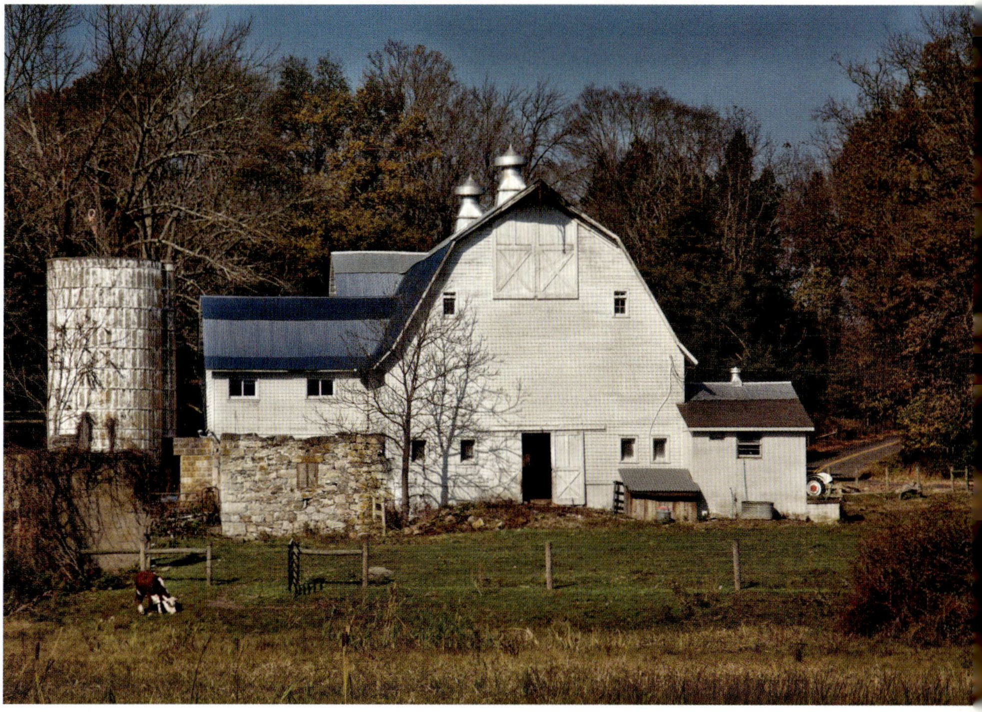

**top**—Kentucky.

**above**—Pennsylvania.

**left**—The dairy farms of Vermont are iconic. They are known for producing butter and cheese.

# Hay Barns

The upper area of a barn was used to dry and store hay. This area is called the mow (rhymes with cow) or the hayloft.

A large door at the top of the ends of the barn could be opened so that hay could be brought in.

**top and bottom—***Misty Morning Harvest.* West Virginia.

**following page—**New Jersey.

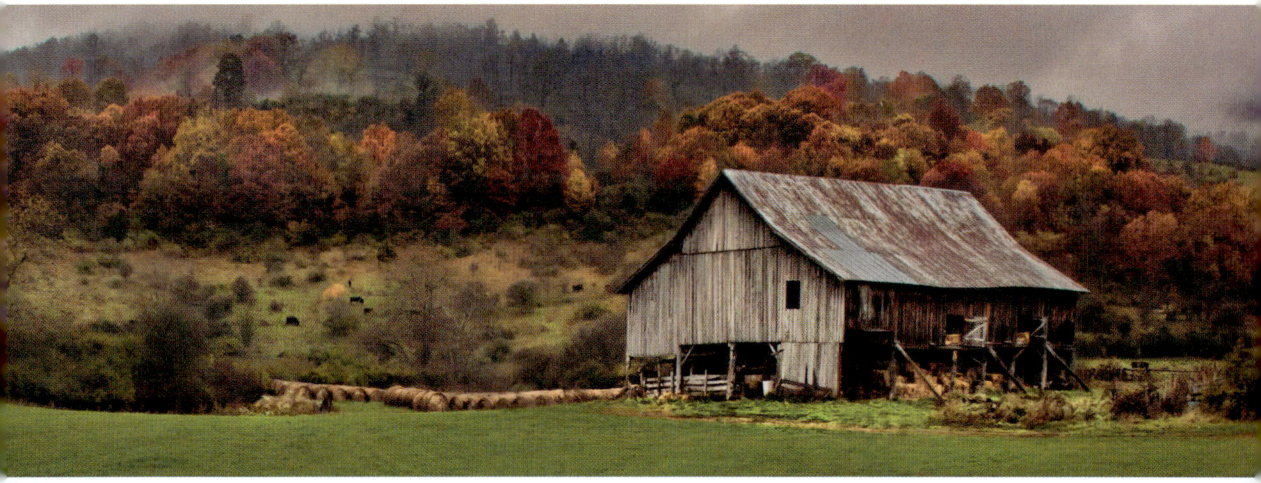

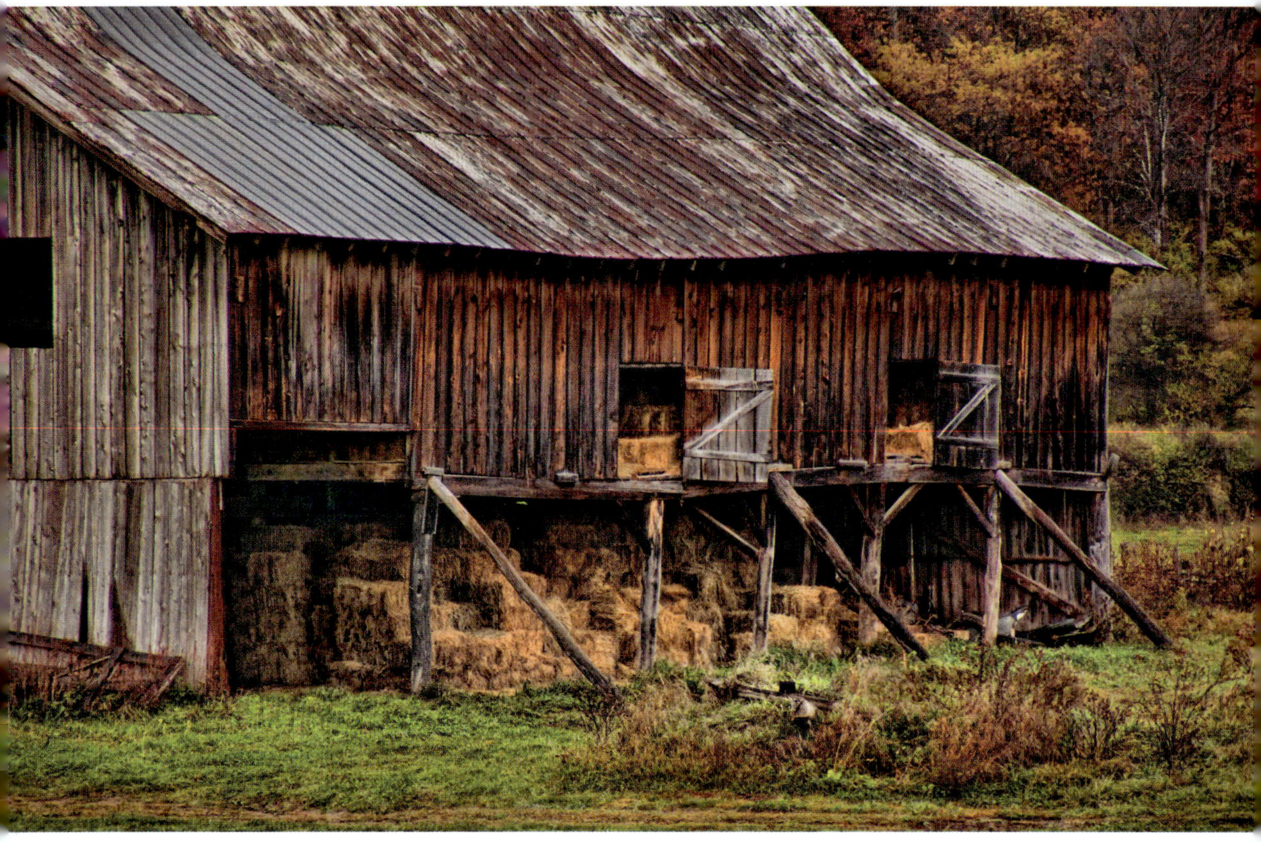

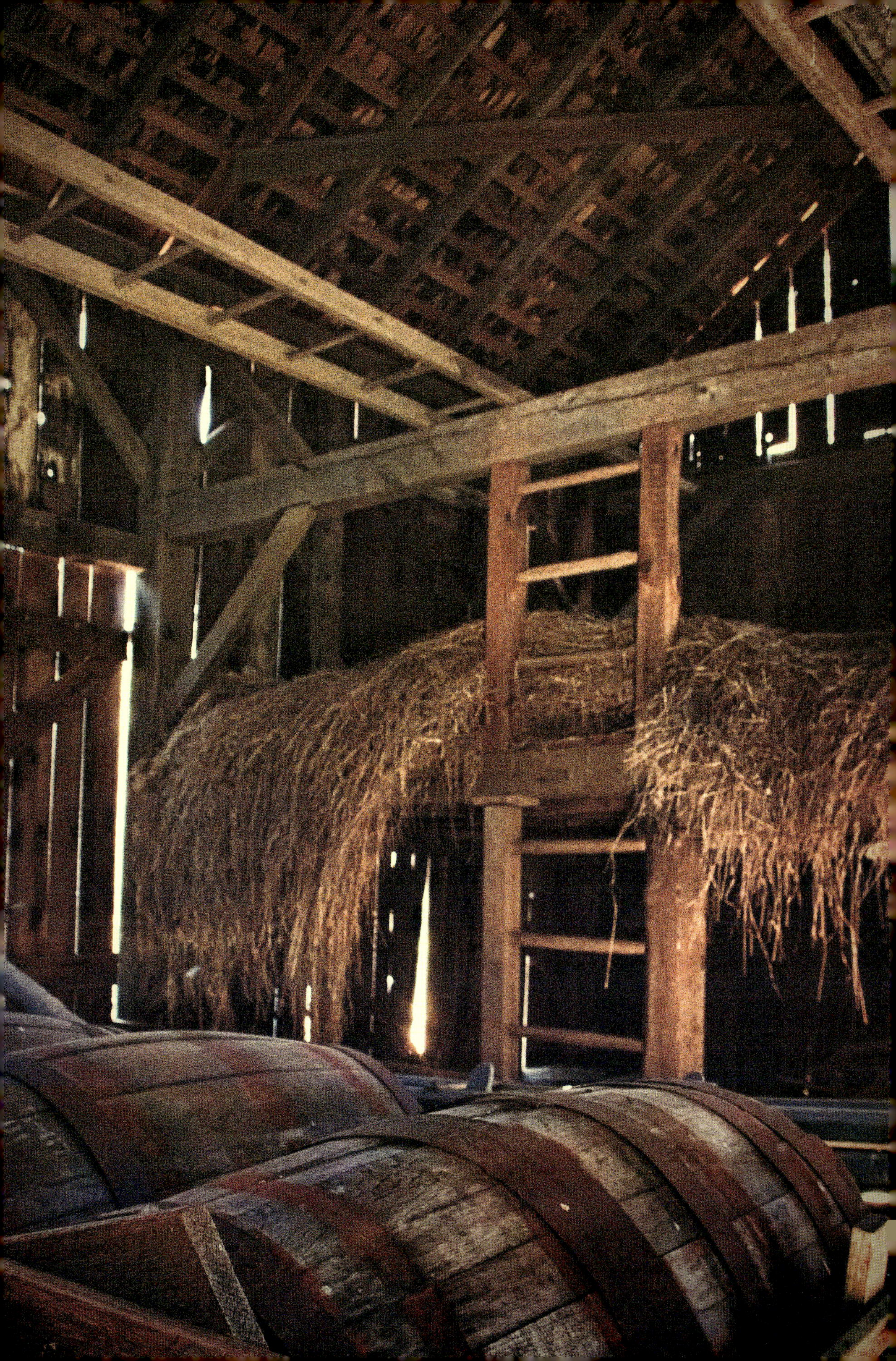

# Components

## Silos

Silos were tall, cylindrical structures that farmers used to store silage (chopped corn), which they fed their cattle during the cold winter months. Temperature and moisture control was critical to prevent contamination of the grain.

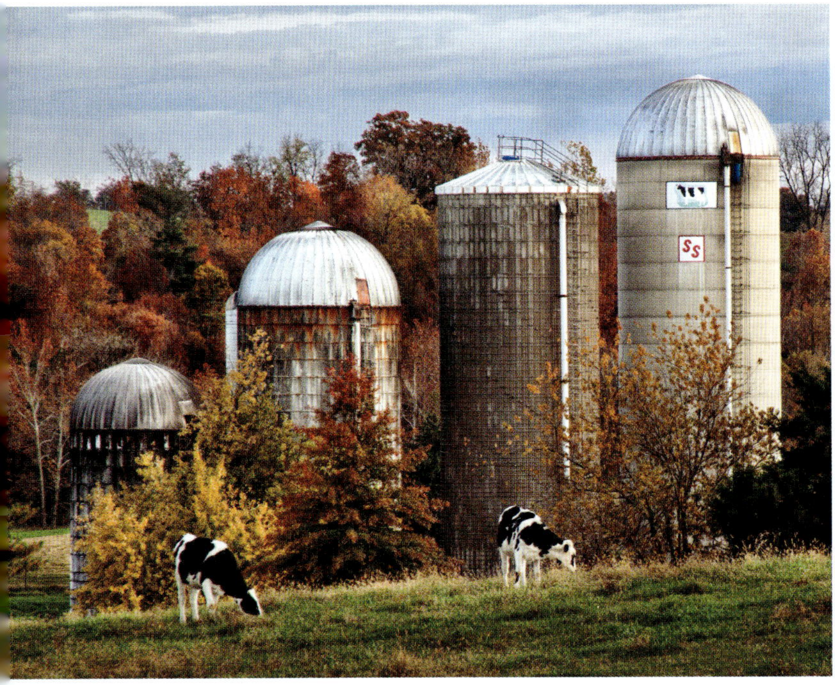

### To the Top

Vertical elevators were used to fill silos and in most cases, gravity was used to empty them. A machine called an auger was used to transport the grain.

**left and below**—Virginia.

**following page**—Pennsylvania.

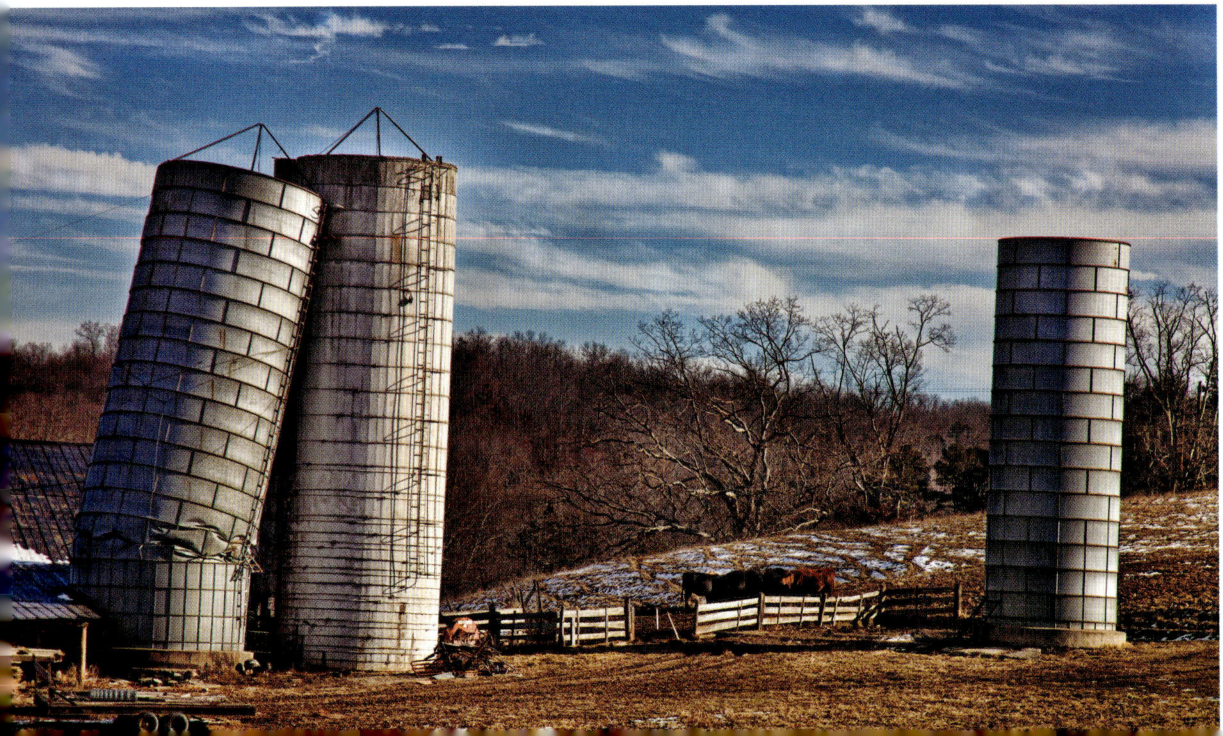

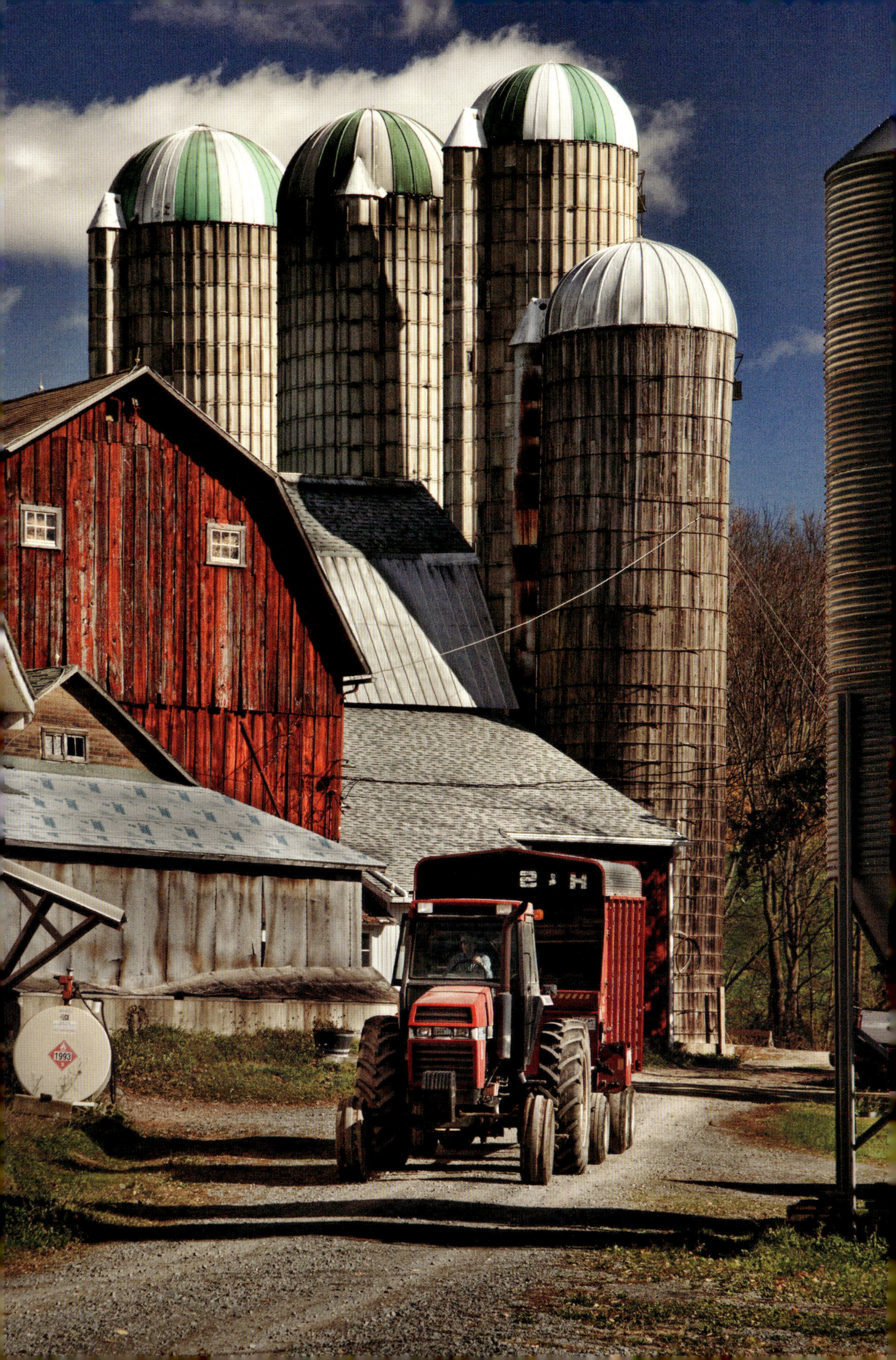
1993

**top left and right**—*Pairs*. The same silos, photographed from two different perspectives. The golden plant in the foreground of the first image is tobacco. New York.

**bottom**—New York.

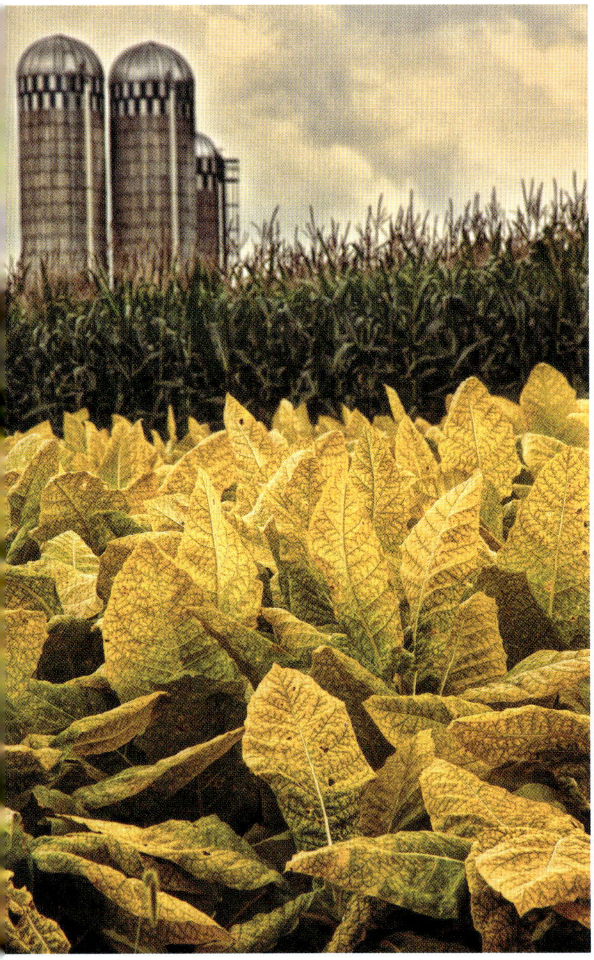

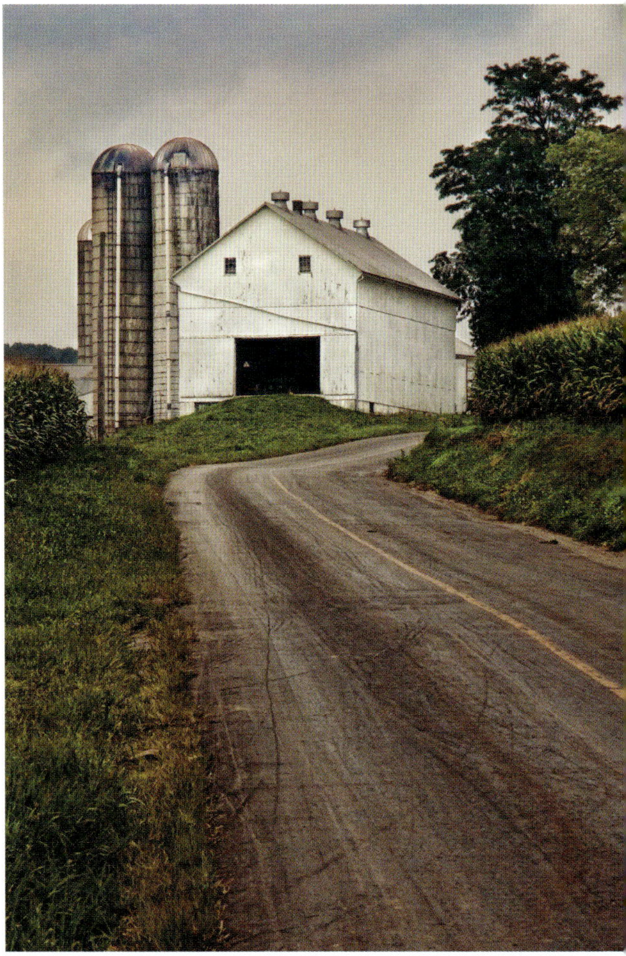

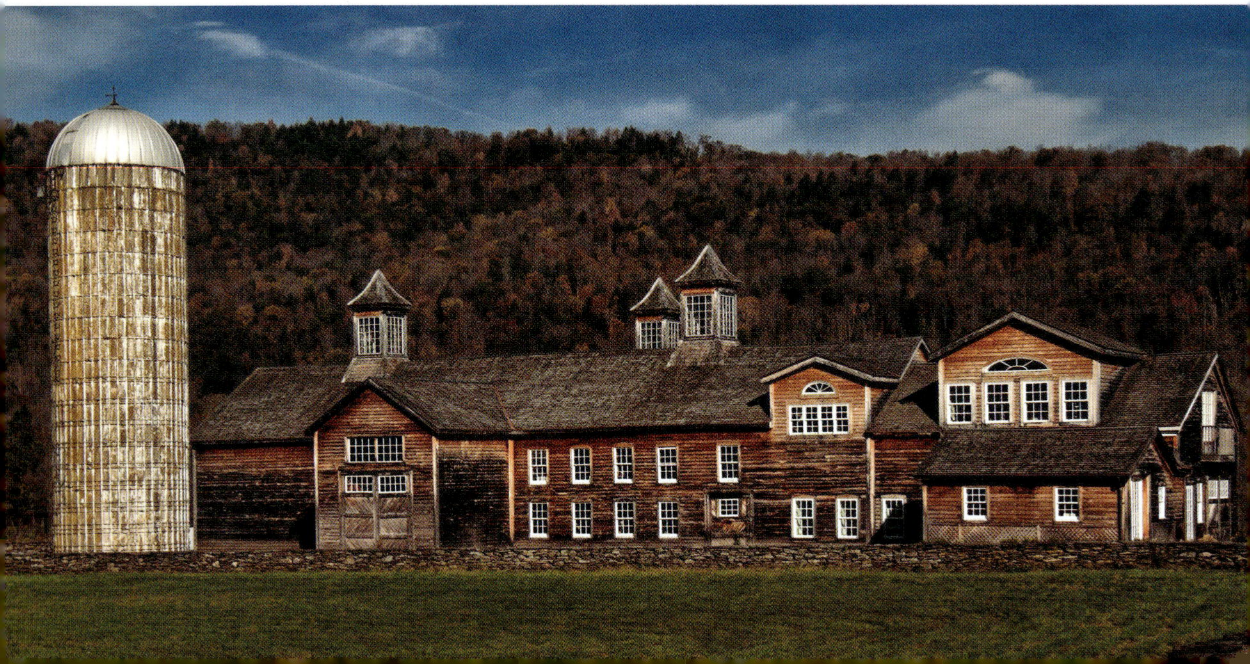

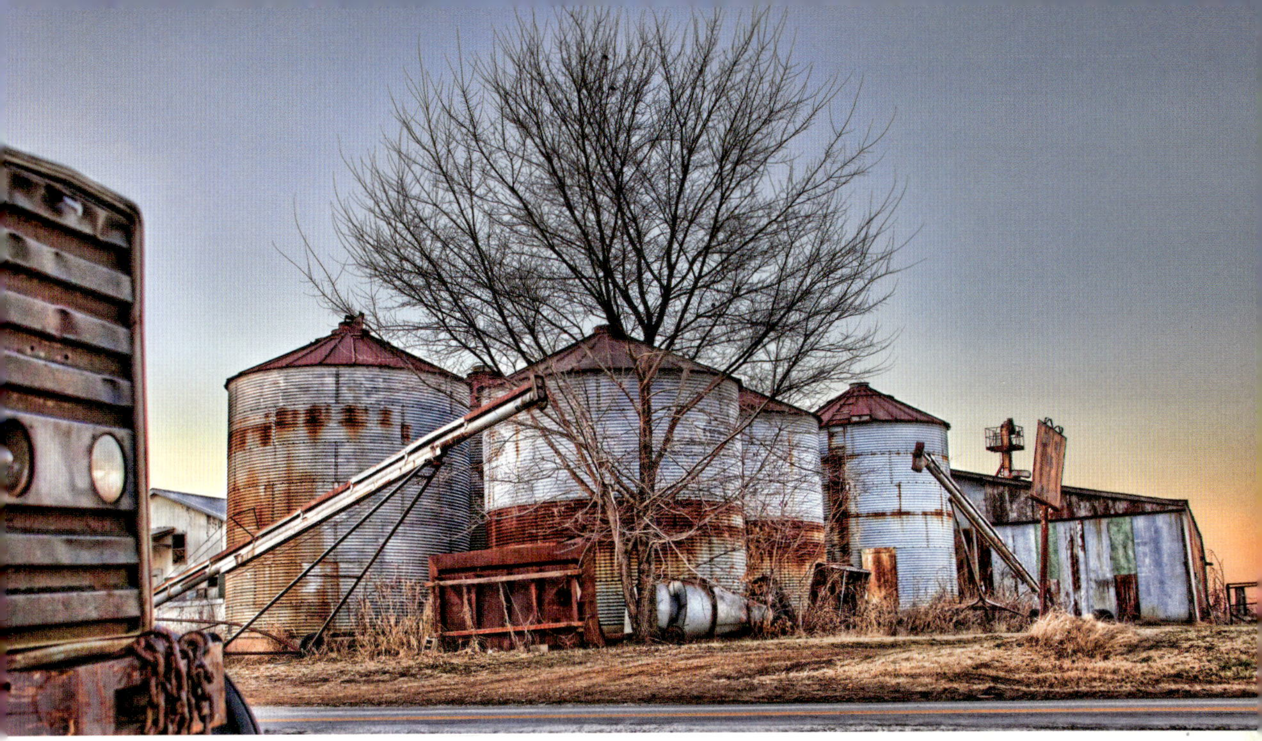

**above**—*Silo City*. Kentucky.

**right**—Massachusetts.

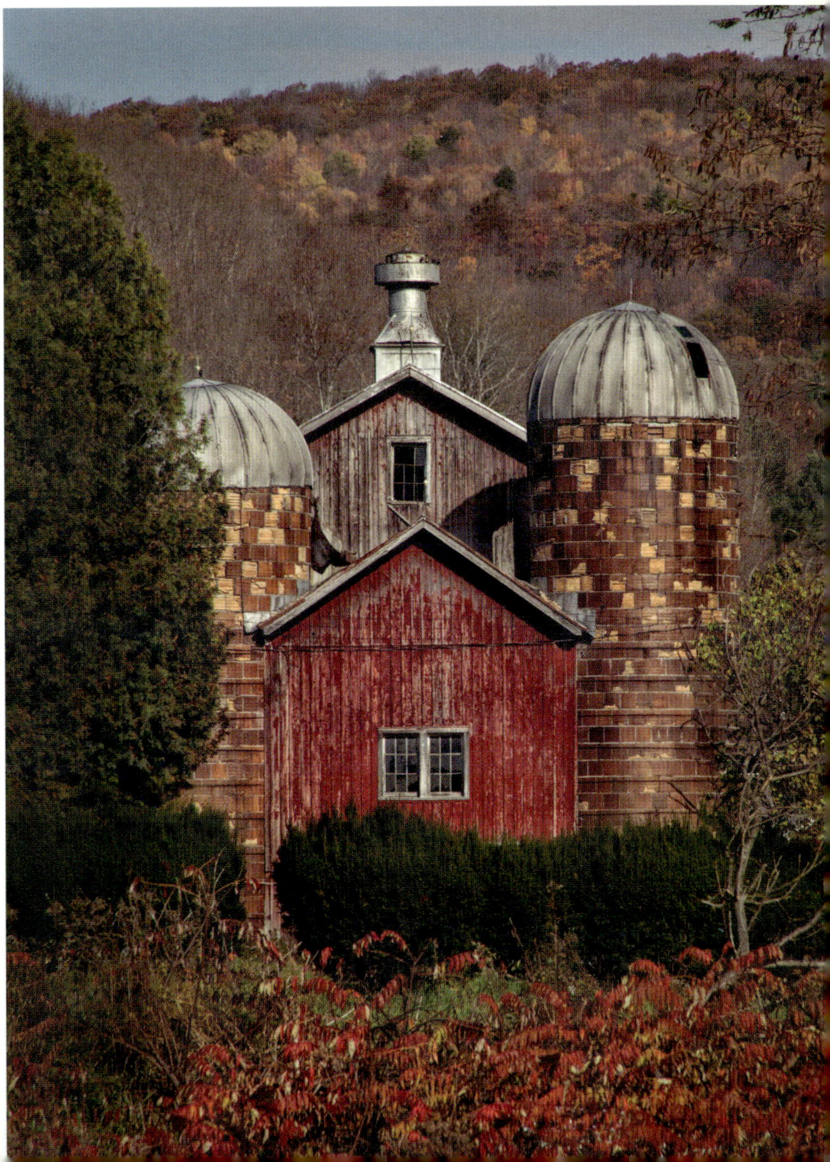

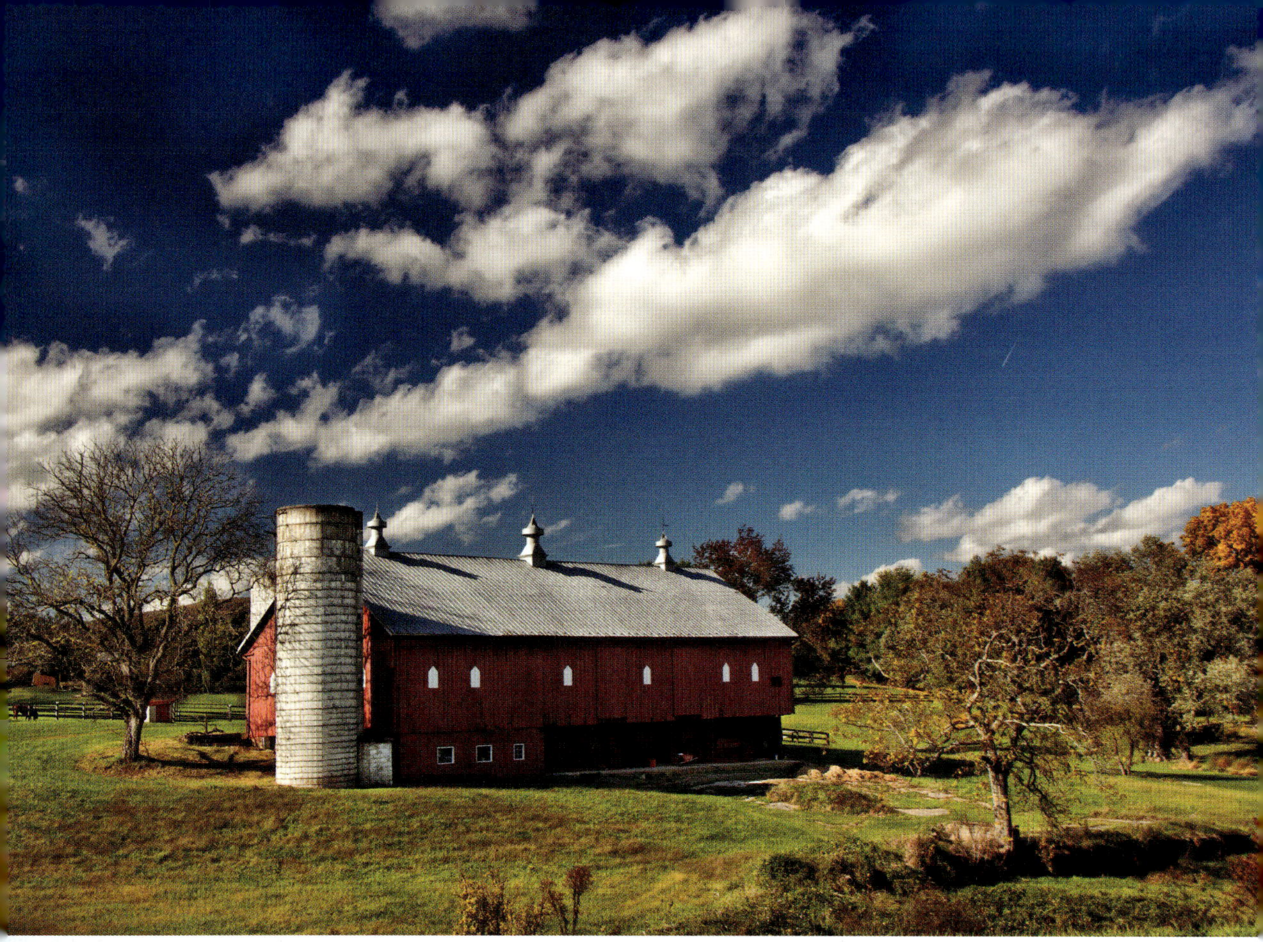

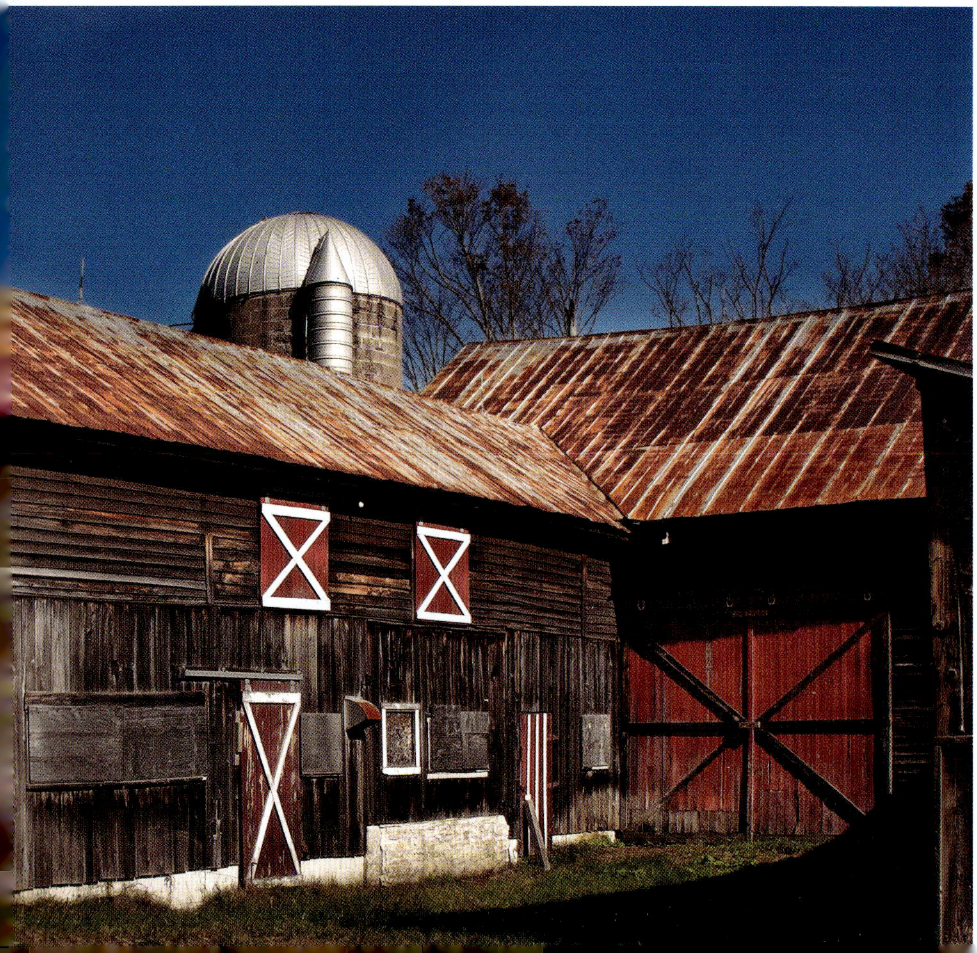

**above**—*The Bank of the Adirondacks.* Catskills, New York.

**left**—New York.

**following page**—Pennsylvania.

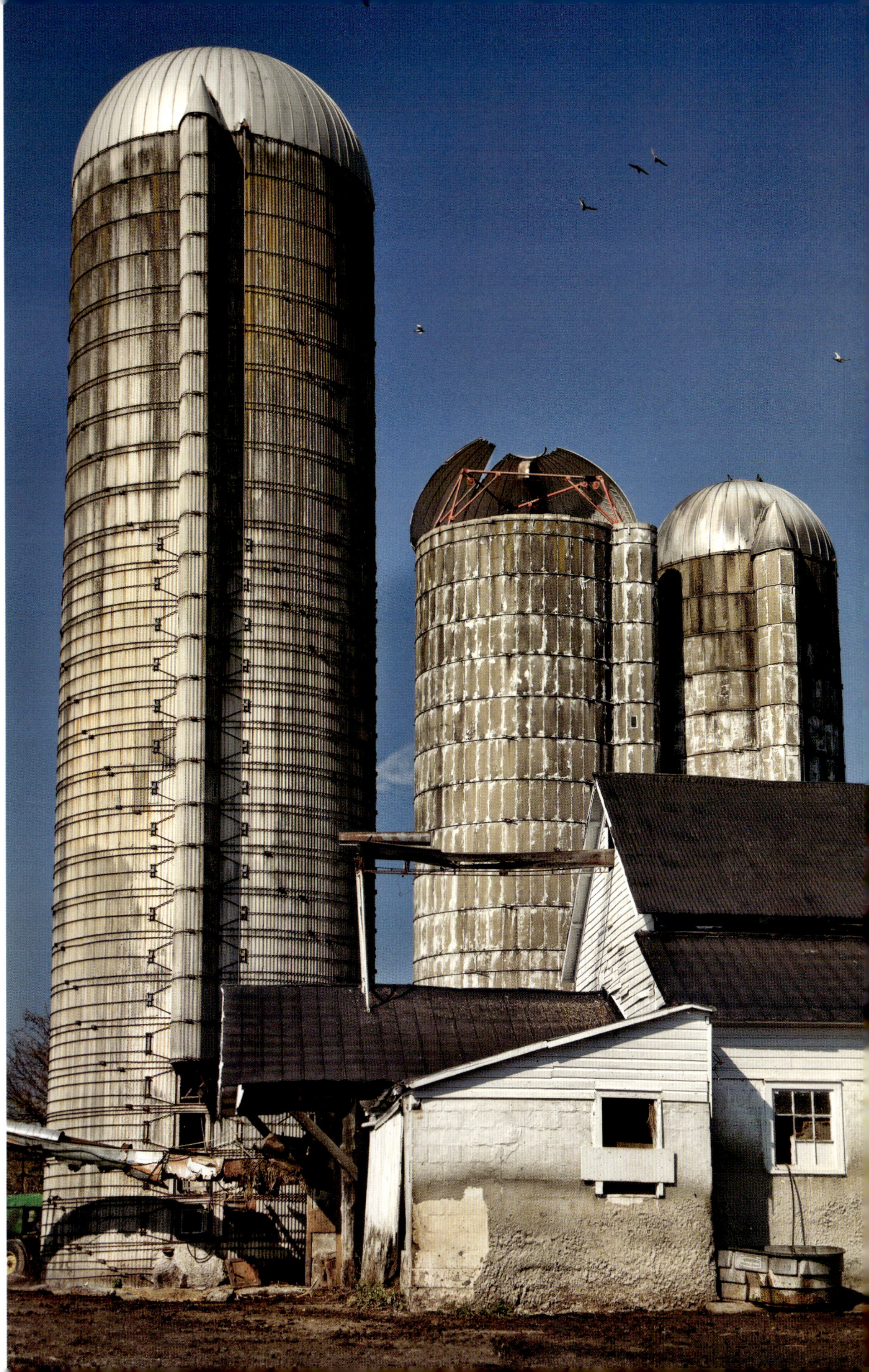

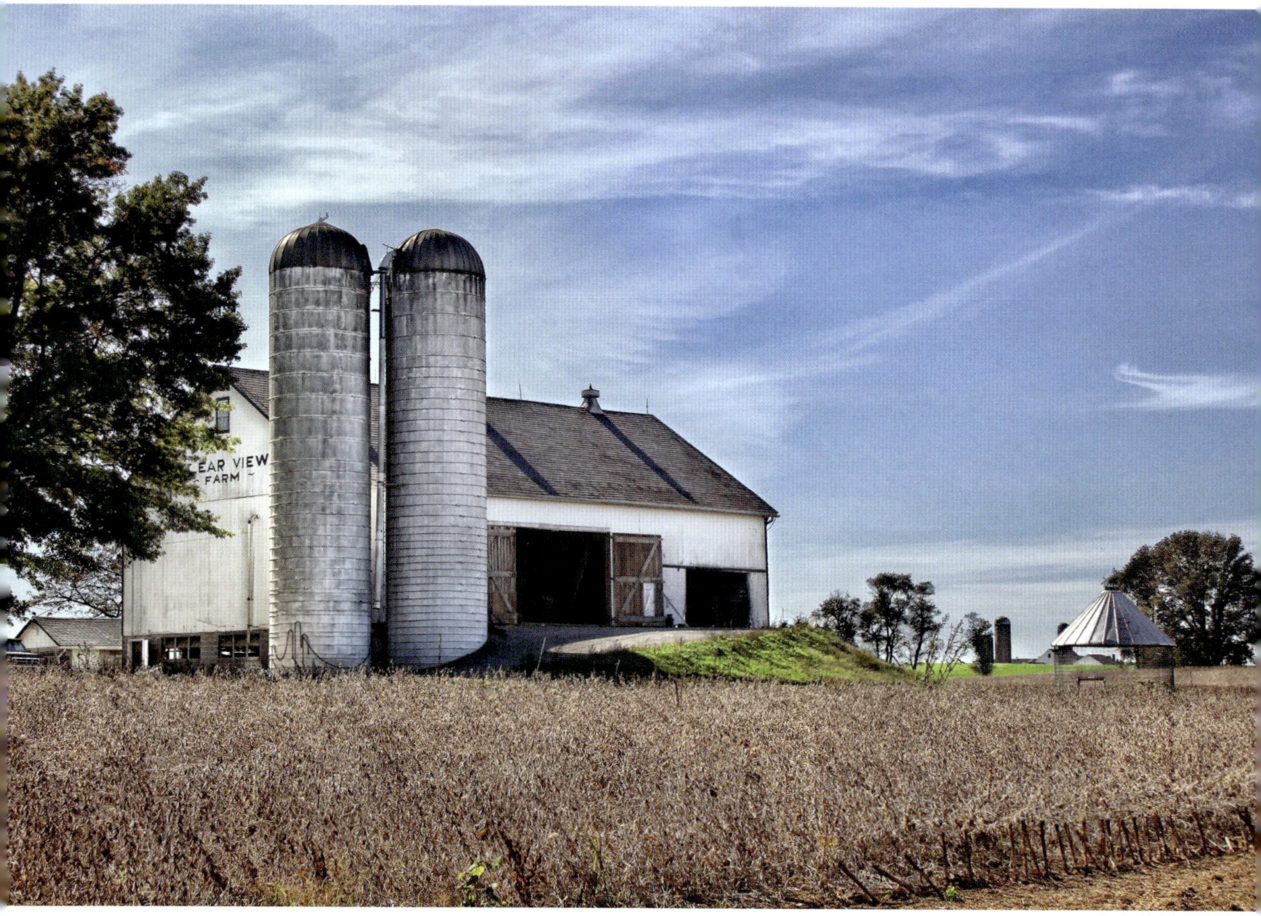
EAR VIEW
FARM

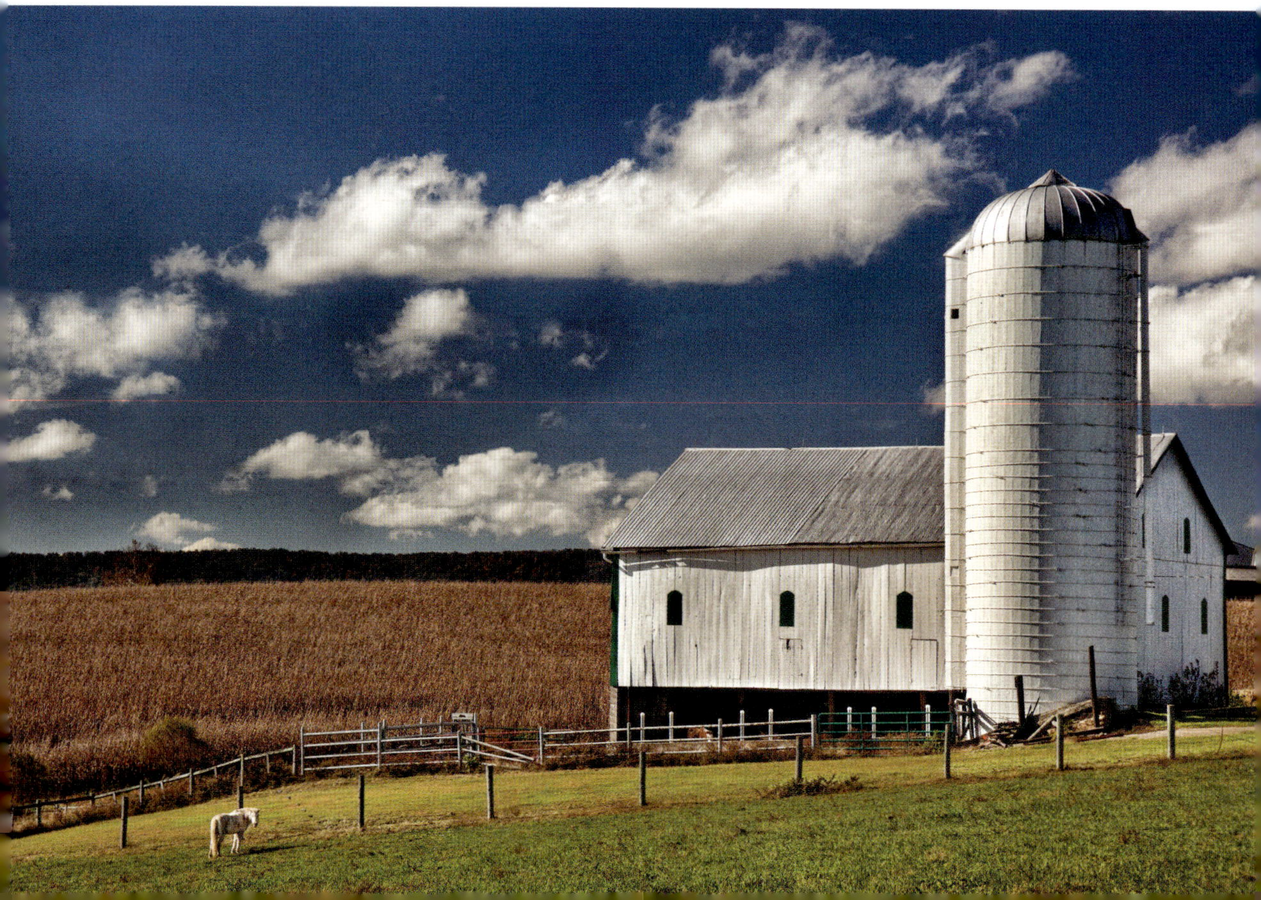

**previous page, top and bottom—**
Pennsylvania.

**right and below—**
New York.

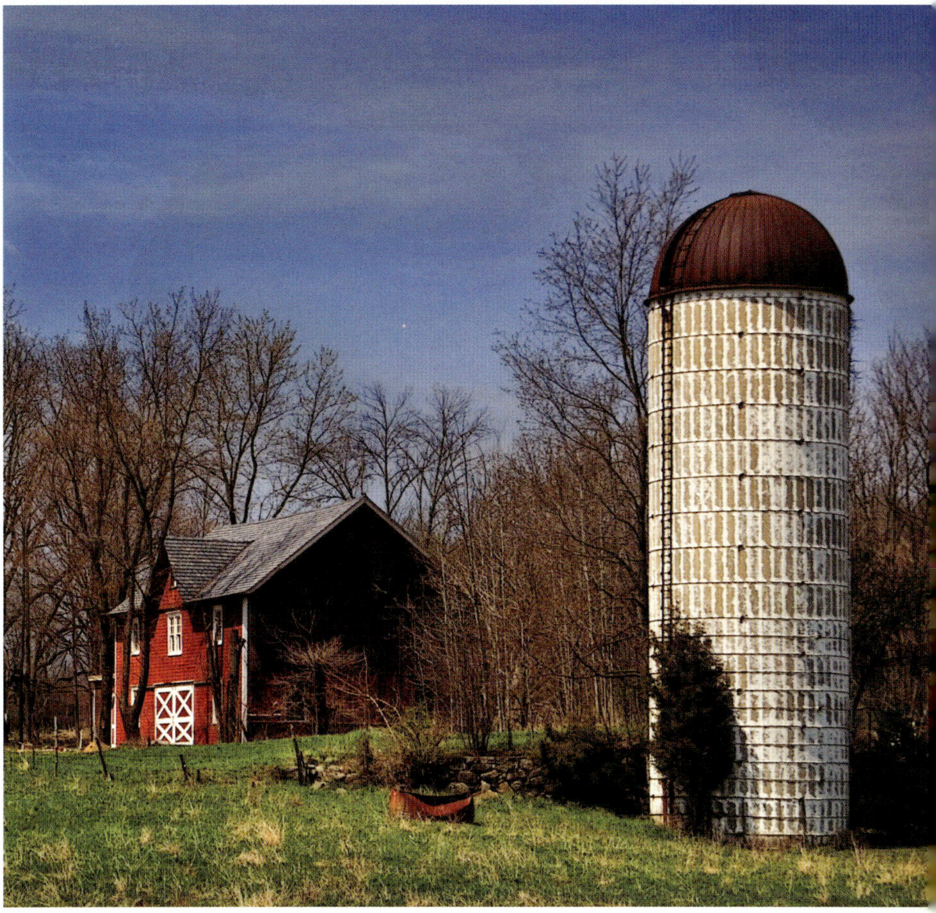

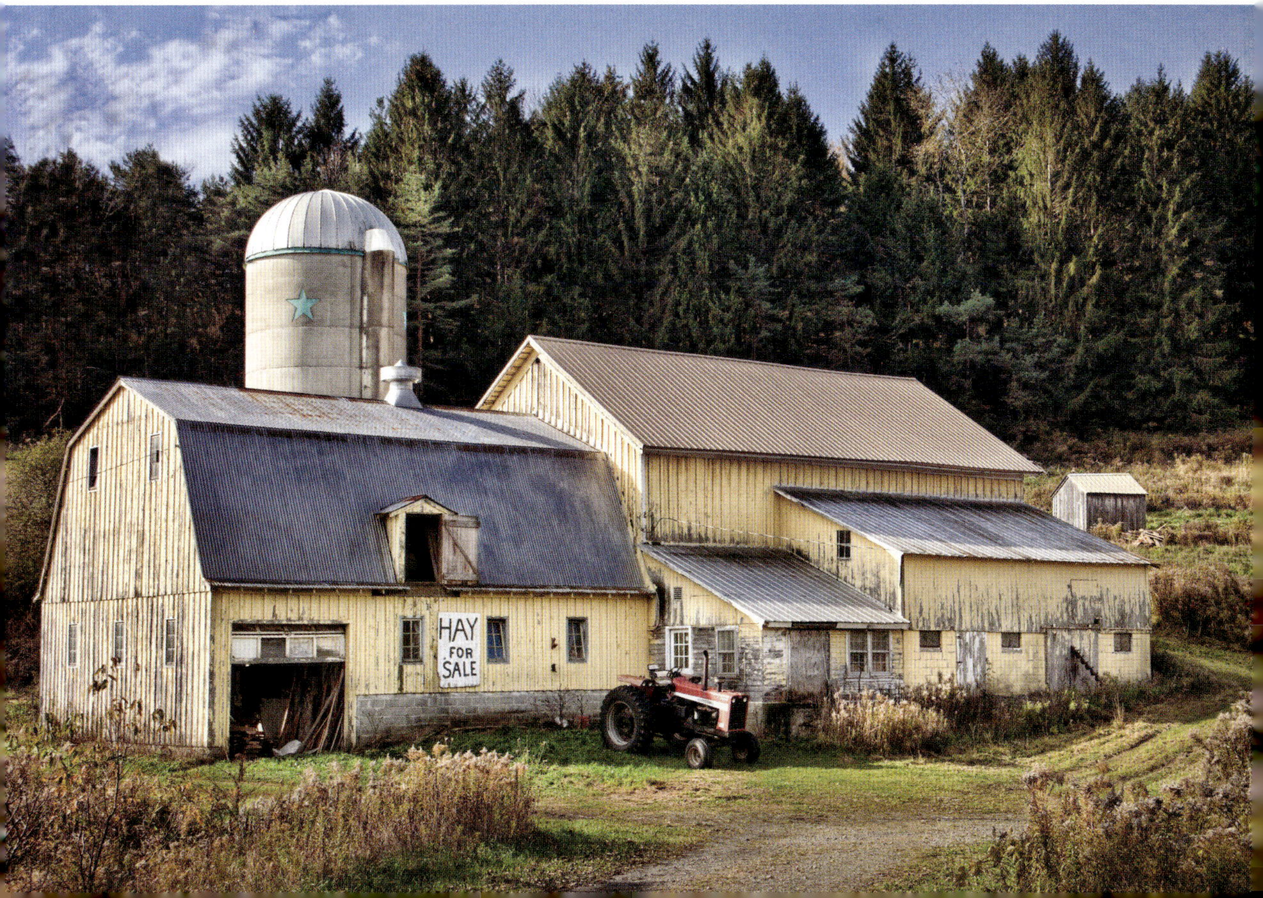

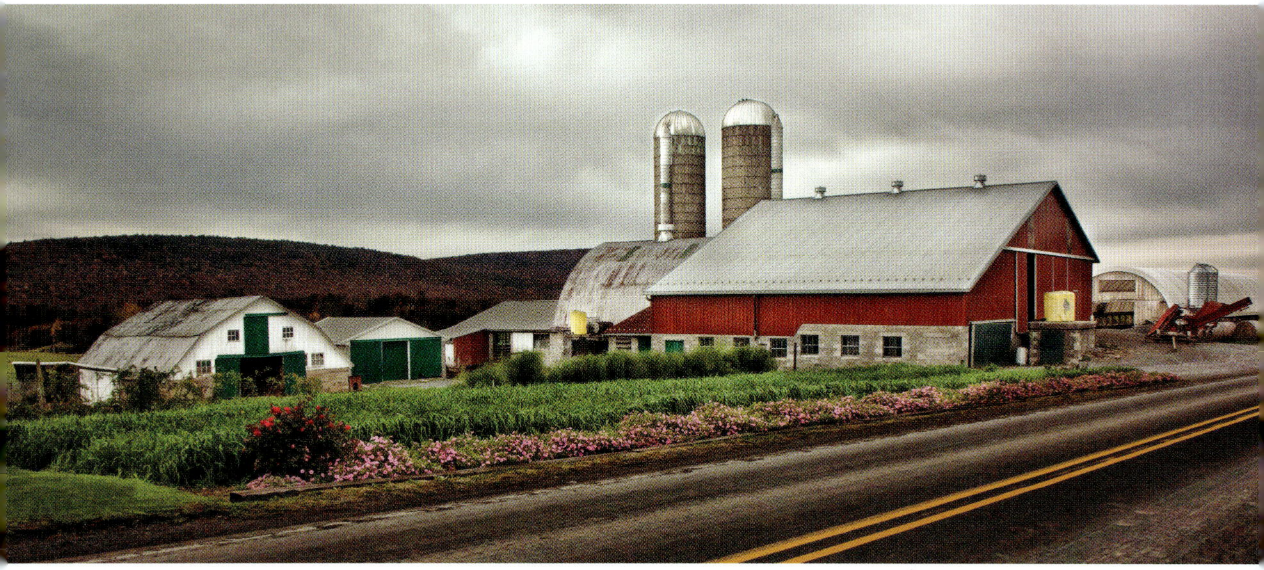

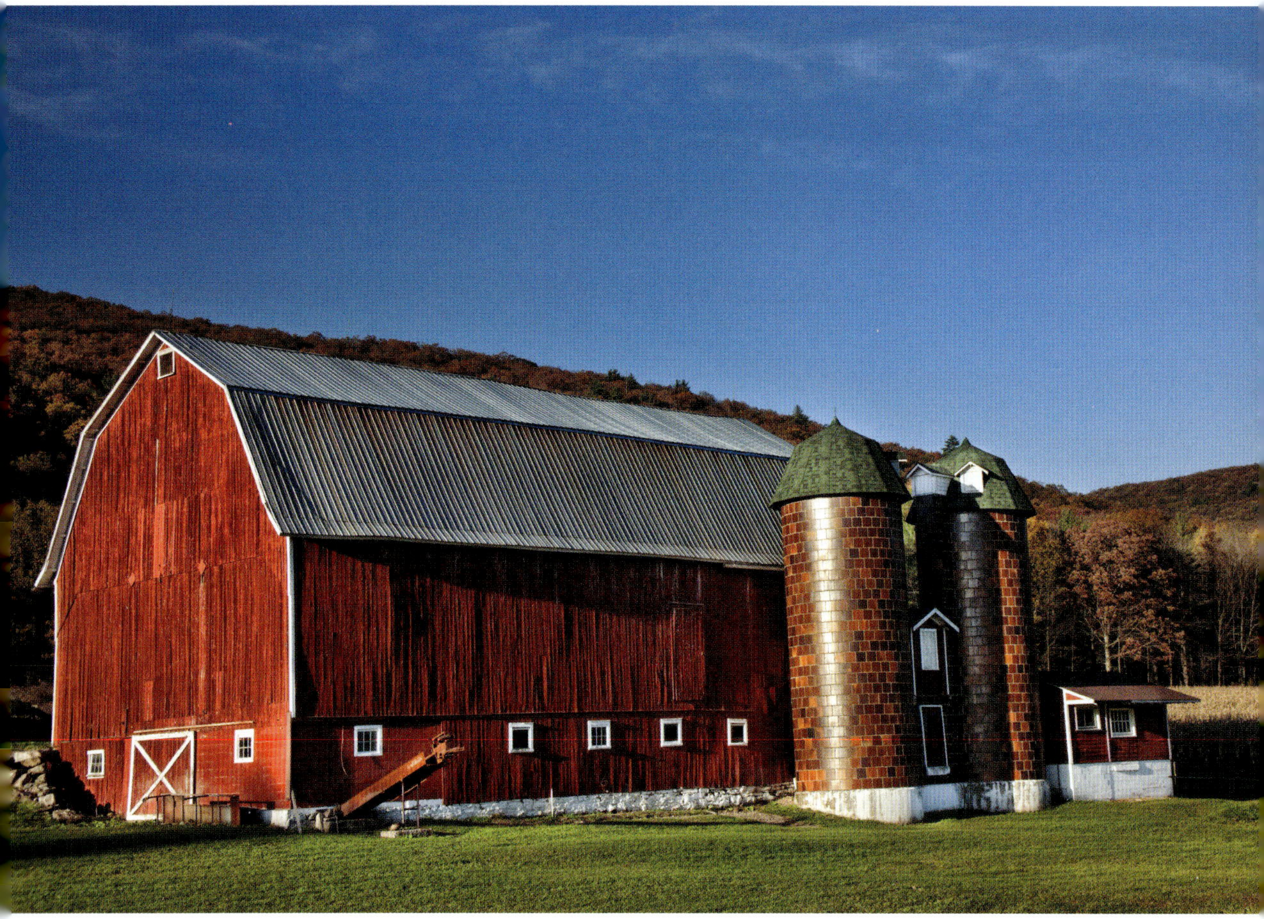

**top**—Pennsylvania.

**bottom**—Massachusetts.

**following page, top and bottom**—New York.

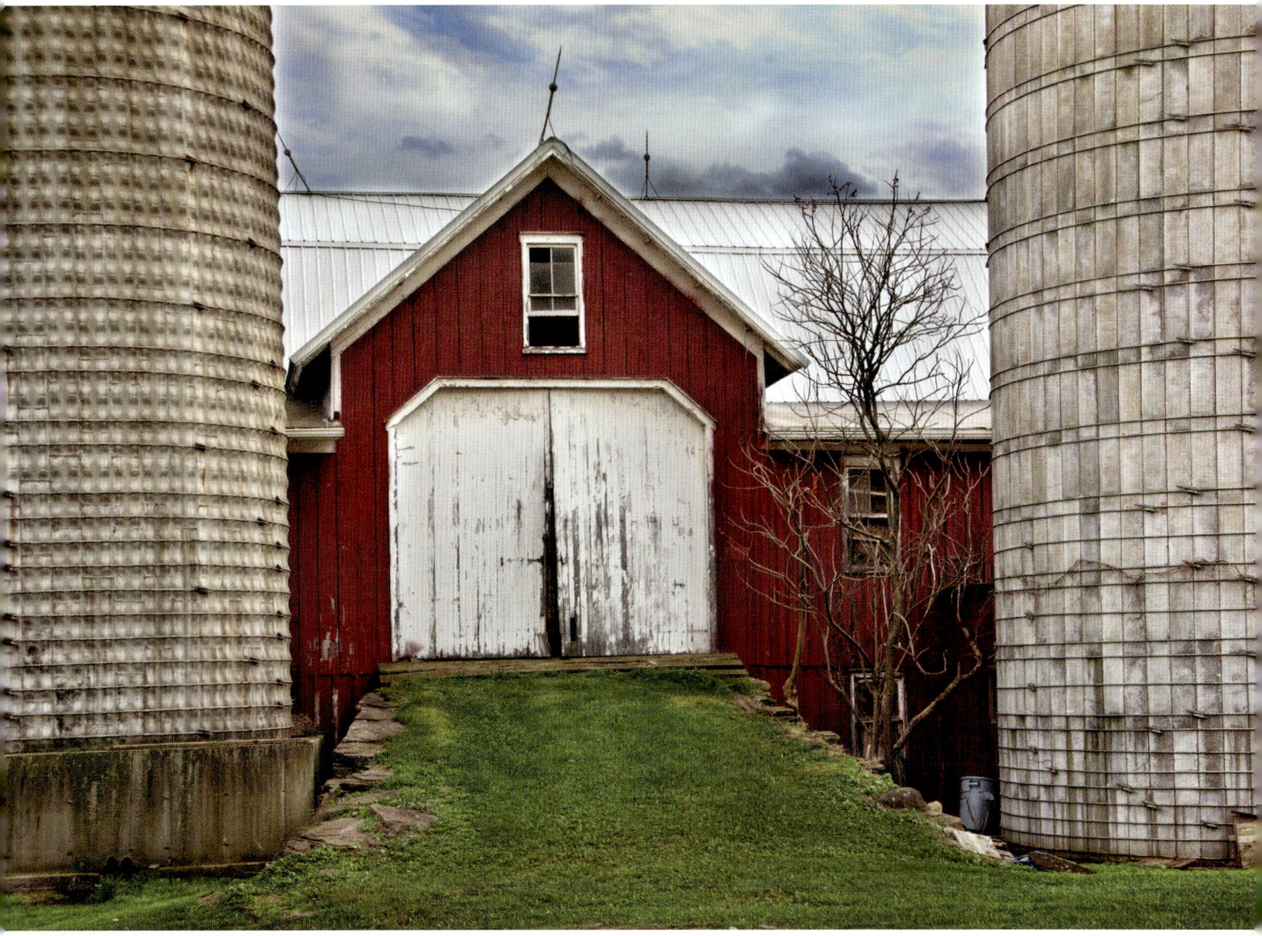

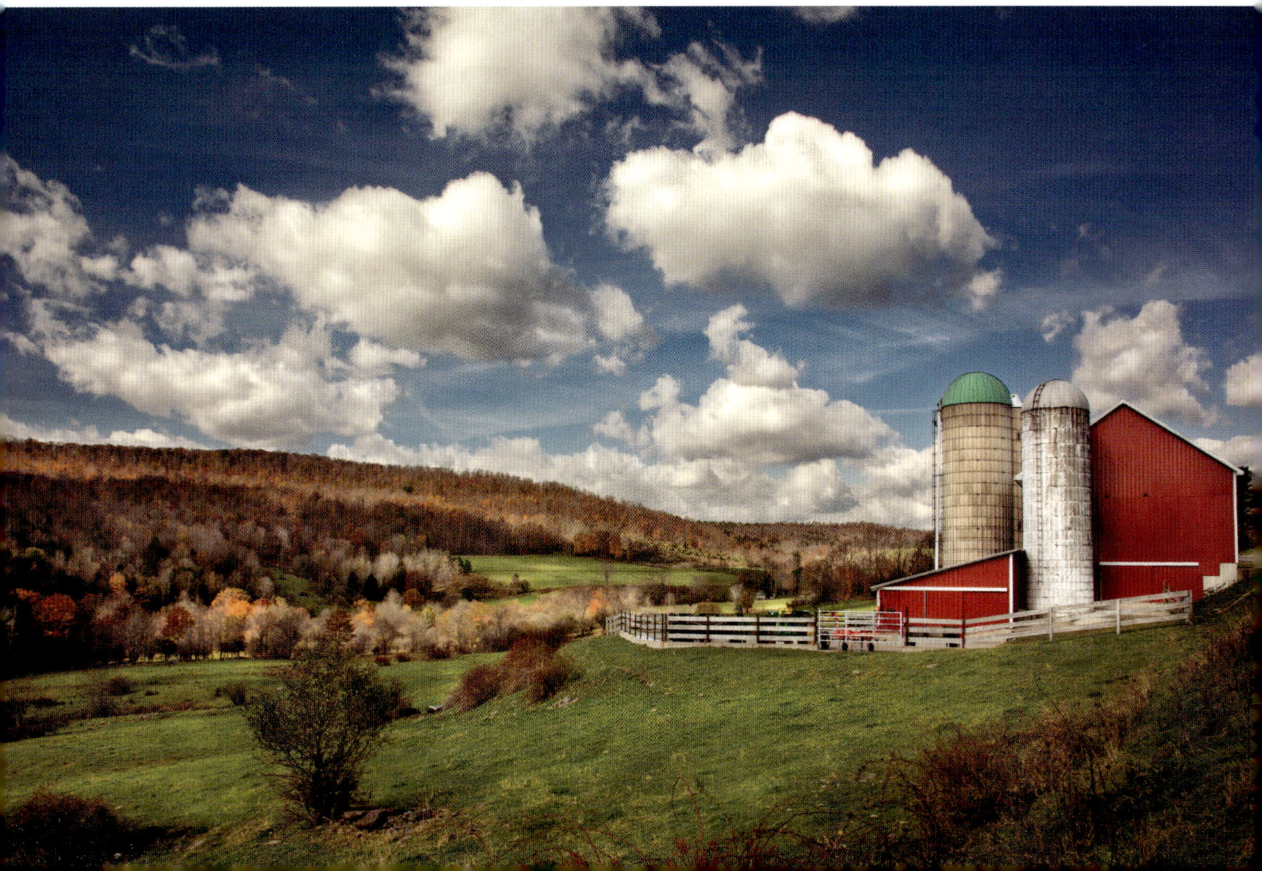

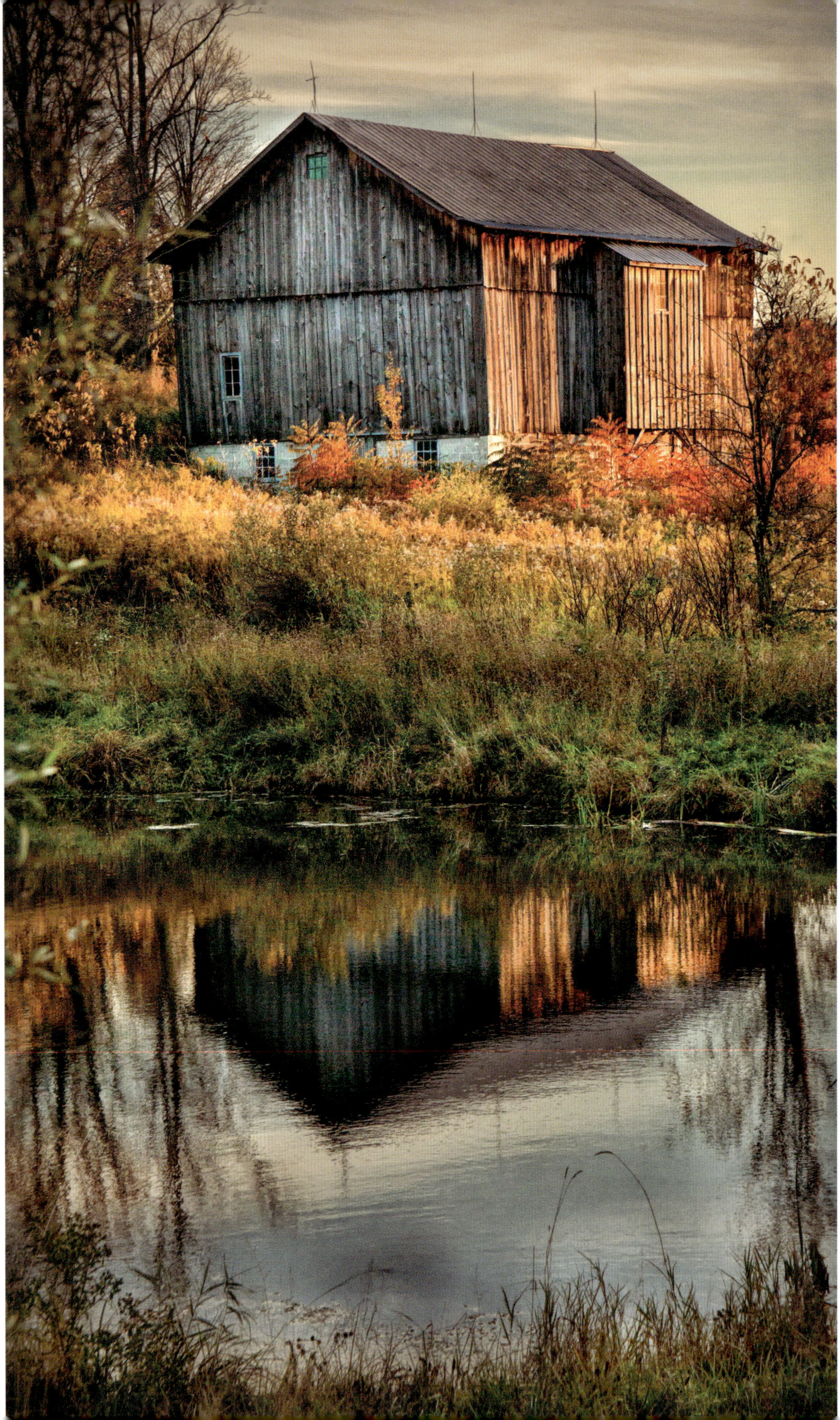

## Lightning Rods

Barns were frequently hit by lighting, so a lighting rod was placed near barns.

## Roofs

Roofs of various styles are commonly used in barn construction. A gambrel is a steep, symmetrical two-sided roof with slopes on both sides, while a gable is a roof with one slope. A salt box roofs is a gable roof built with a lean to one side, usually two stories in the front and one in the back. A Dutch roof is one with both sides extending below the first floor. A Gothic roof has an arch shape and center peak.

**previous page**—Barn with lightning rod. Pennsylvania.

**top right**—Gable roof line. New York.

**below**—A delapidated gable roof. Nebraska.

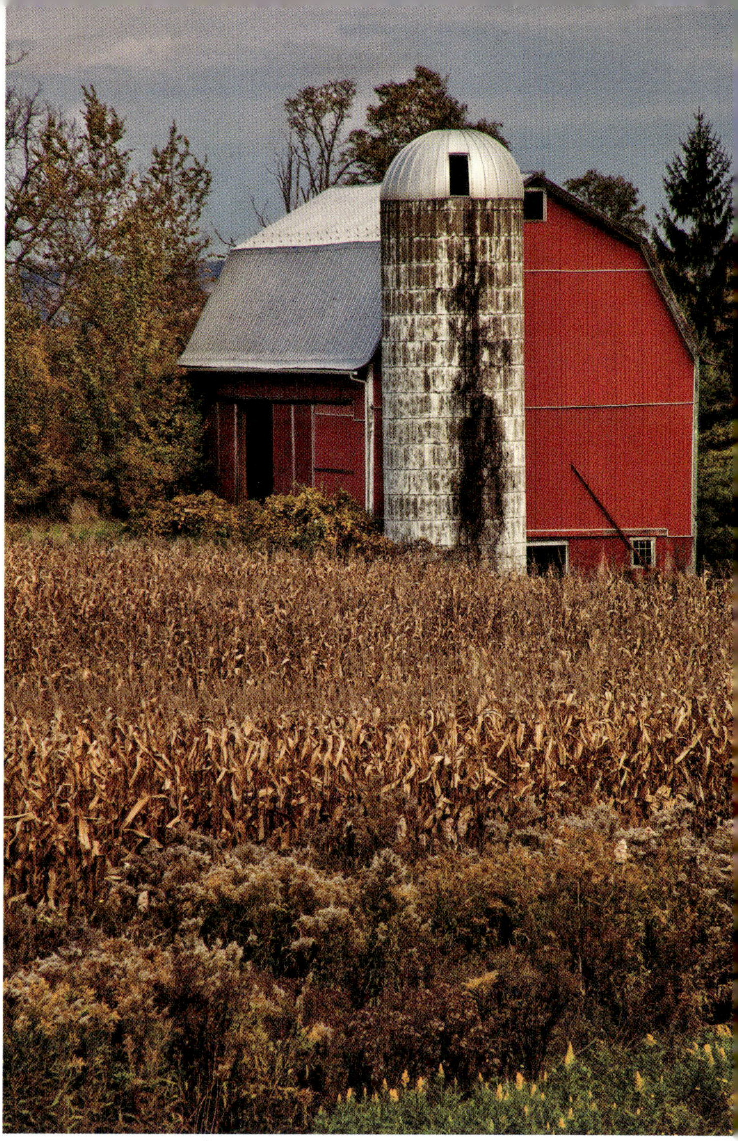

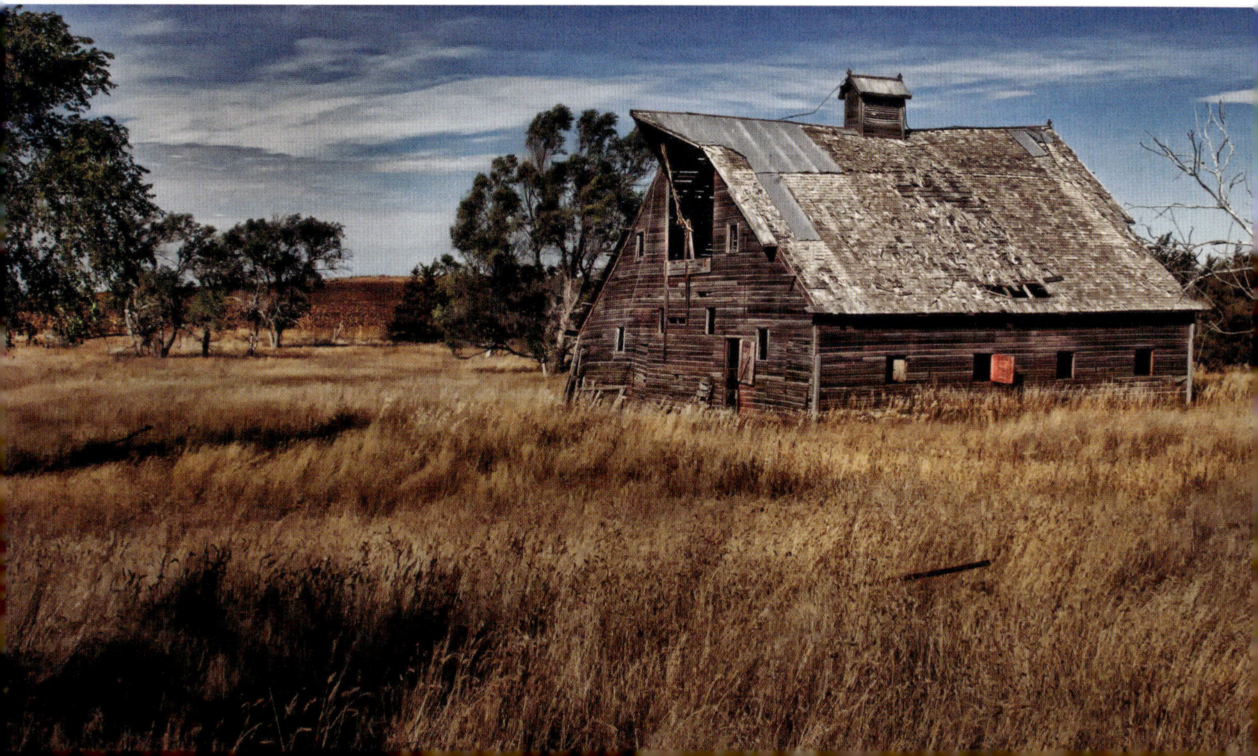

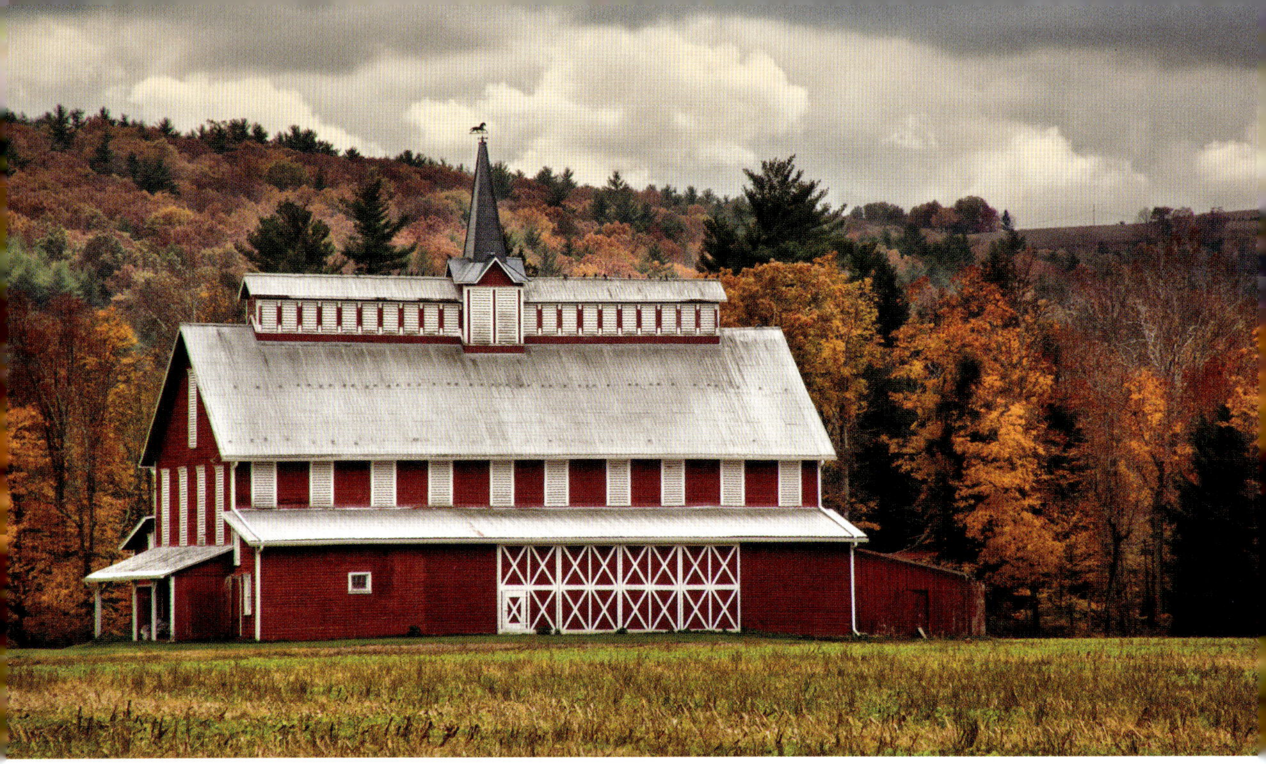

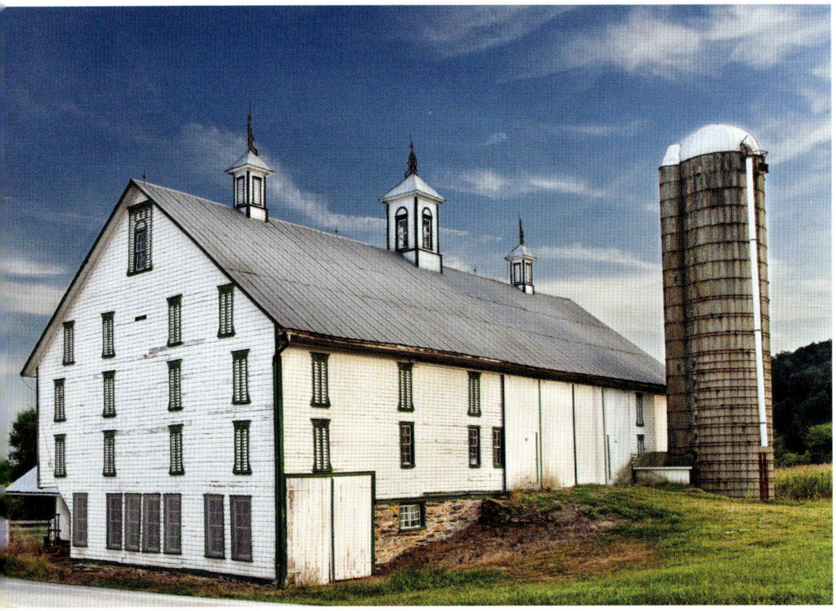

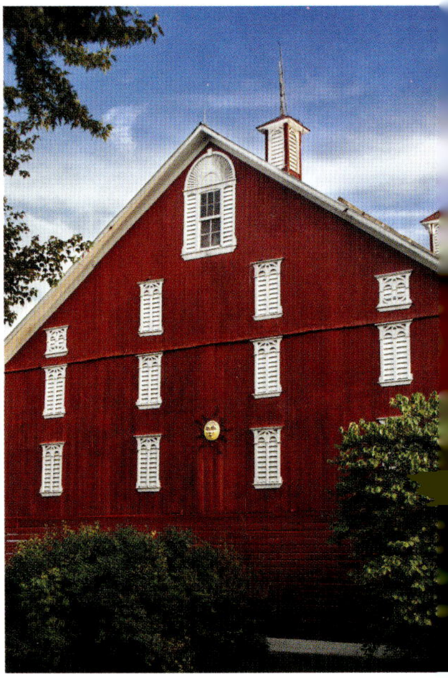

## Cupolas

A cupola is a small tower or dome-like feature projecting from the top of a roof. They were added to provide ventilation and a decorative element to the barn.

**this page**—Three barns with cupolas. Pennsylvania.

**following page**—Allendale, New Jersey.

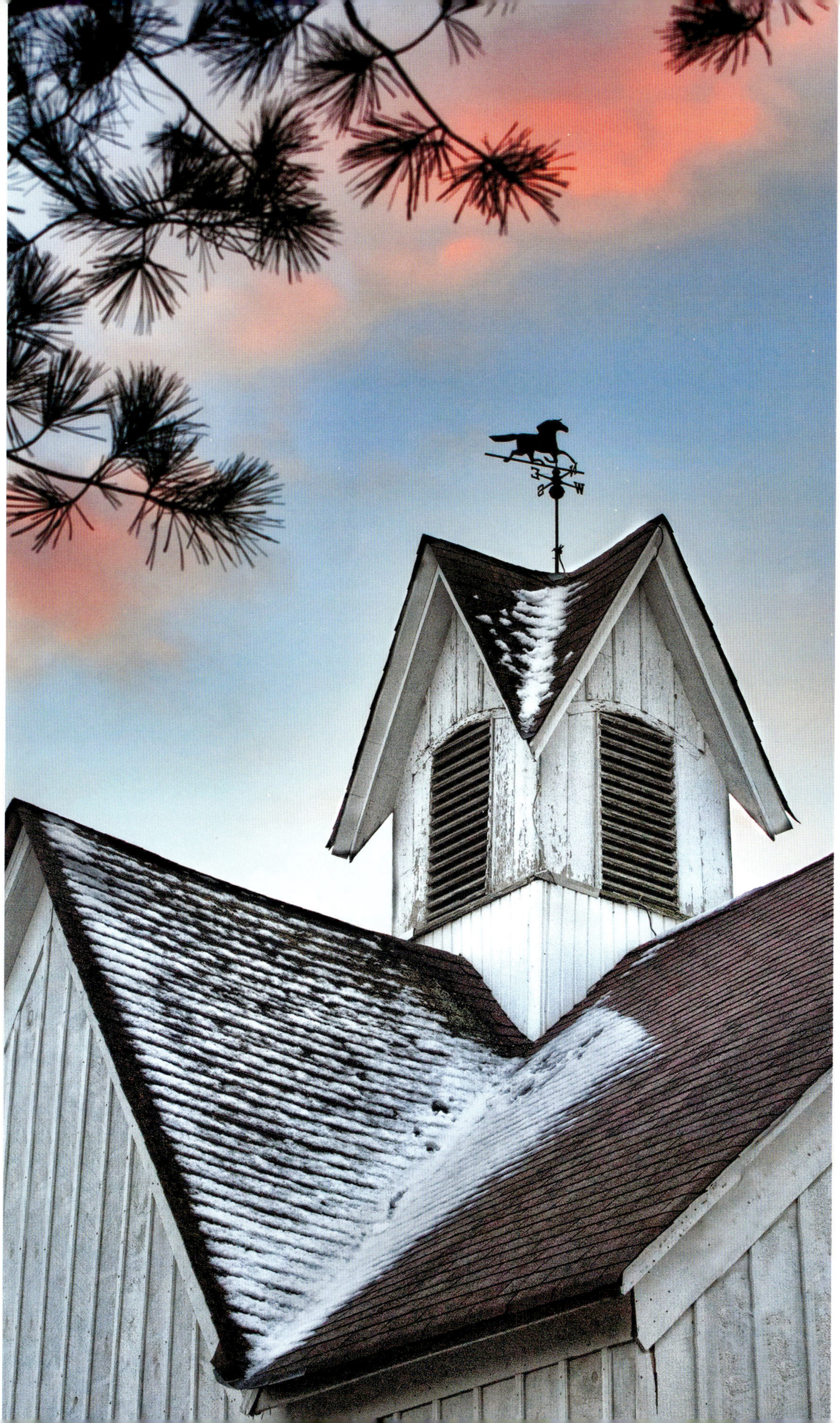

# Doors

Doors were another thoughtfully planned part of barn construction. Sliding doors became popular because farmers could open and close them quickly to keep the cold air out.

**below**—Vermont.

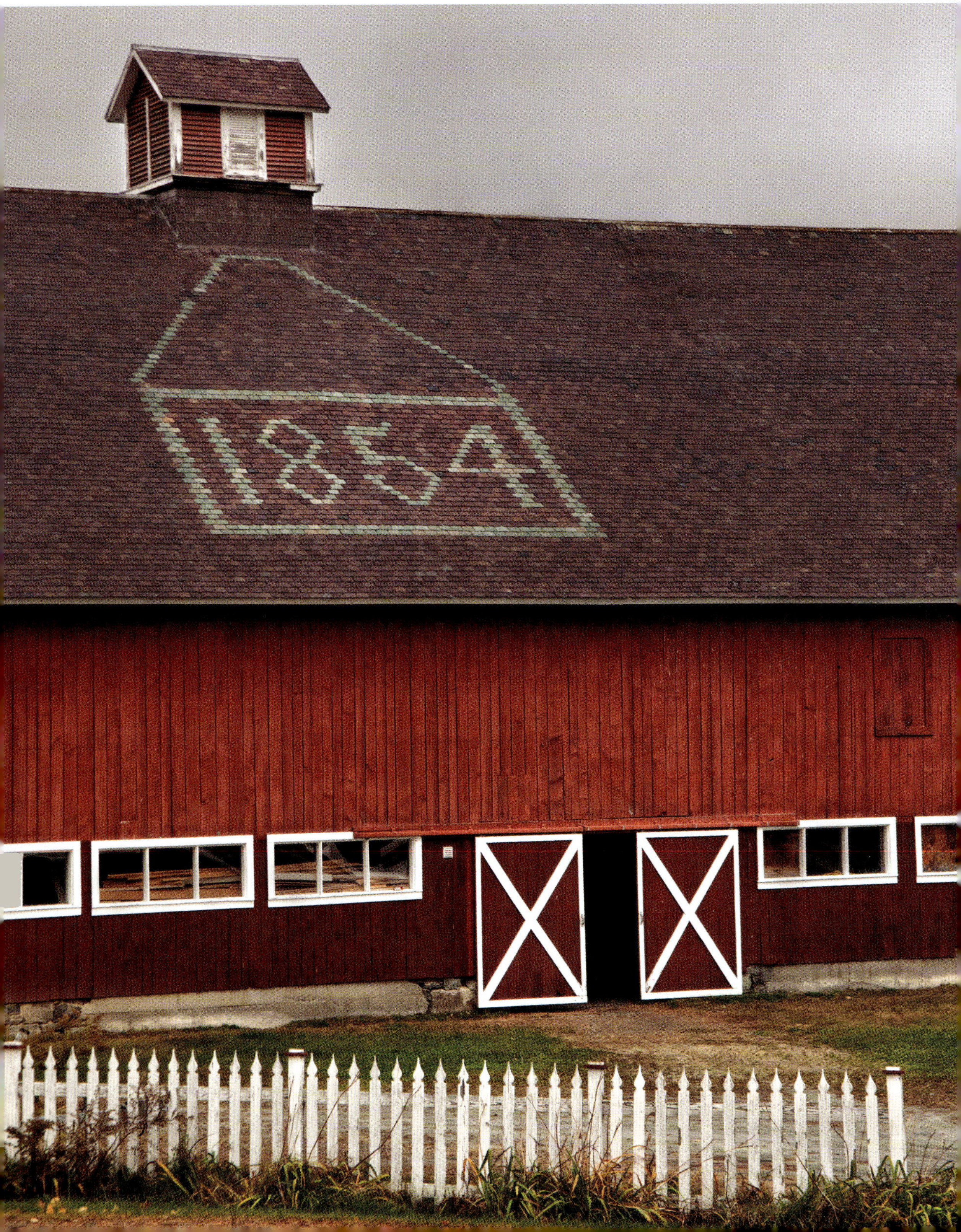

**top**—New York.

**center**—New Hampshire.

**bottom**—Colorado.

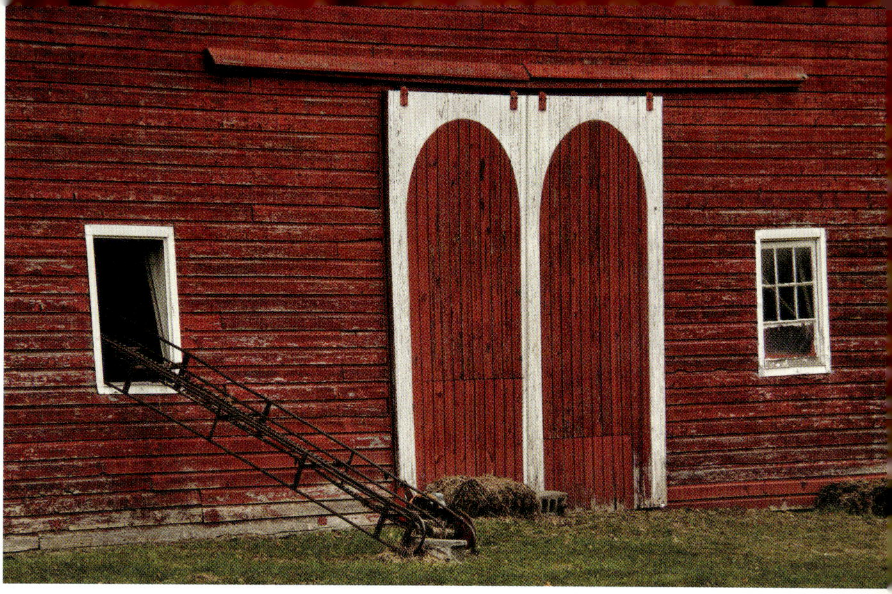

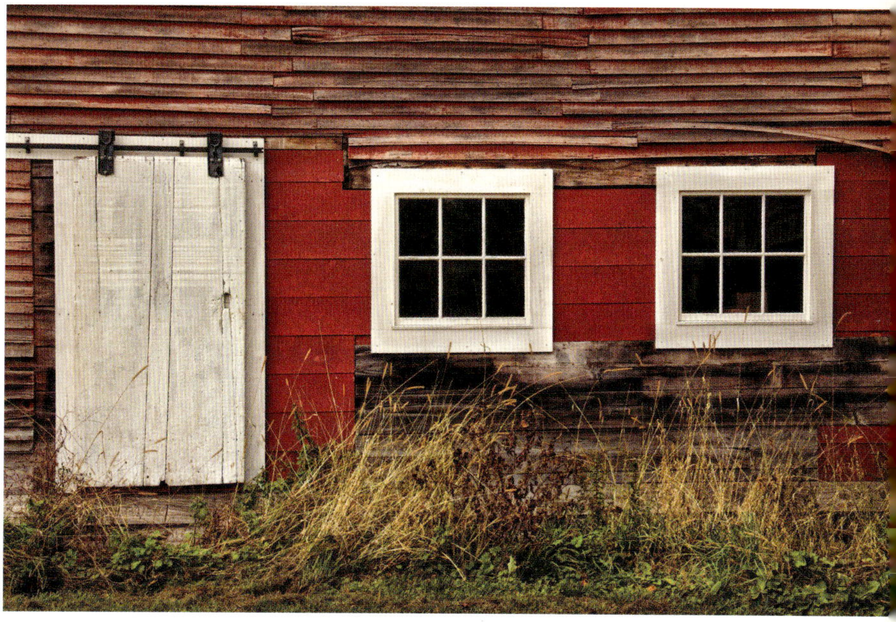

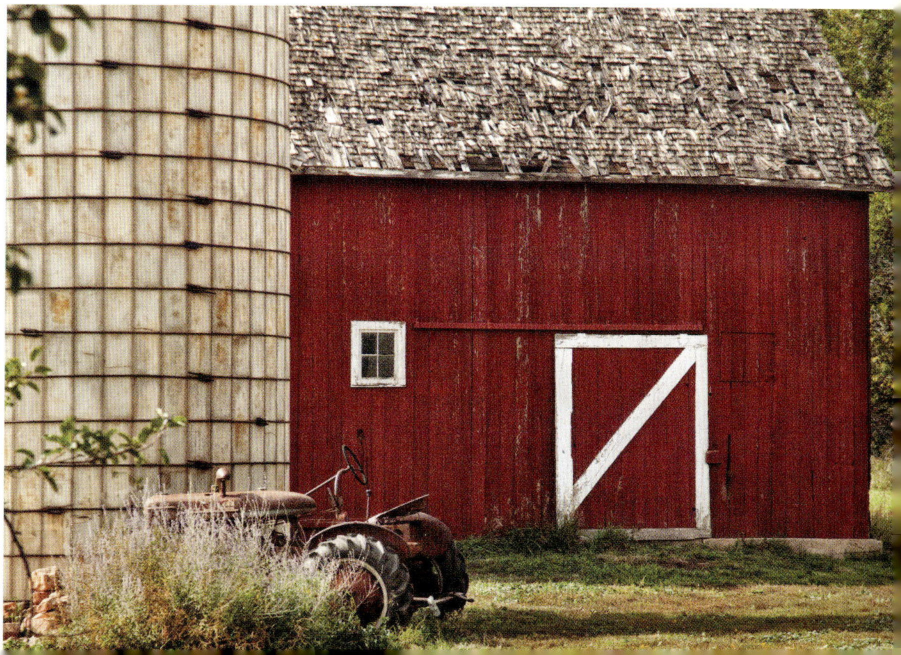

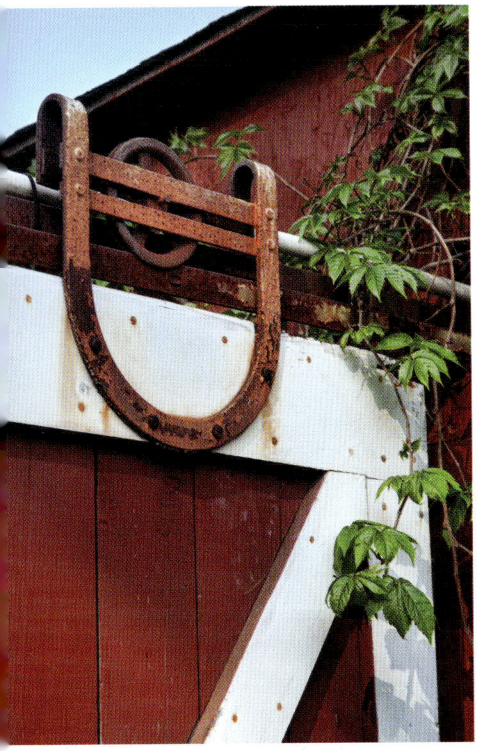

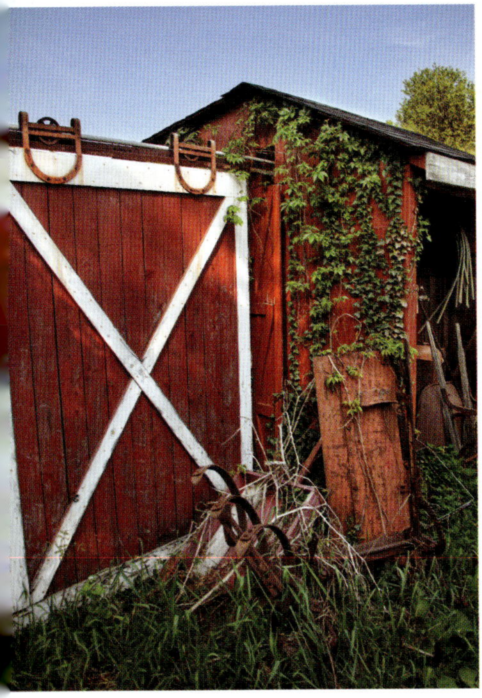

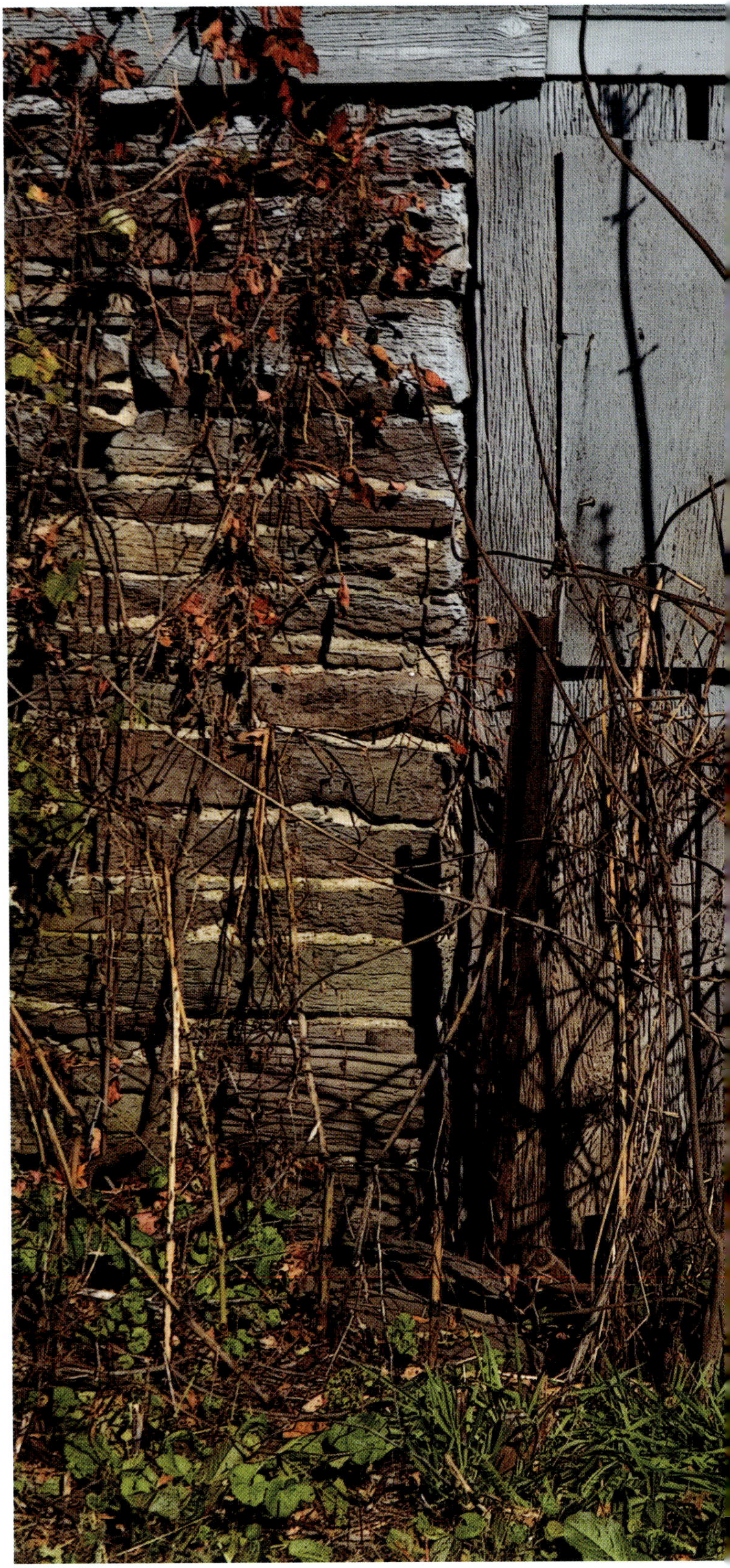

**top and bottom left**—Saddle River, New Jersey.

**right**—Massachusetts.

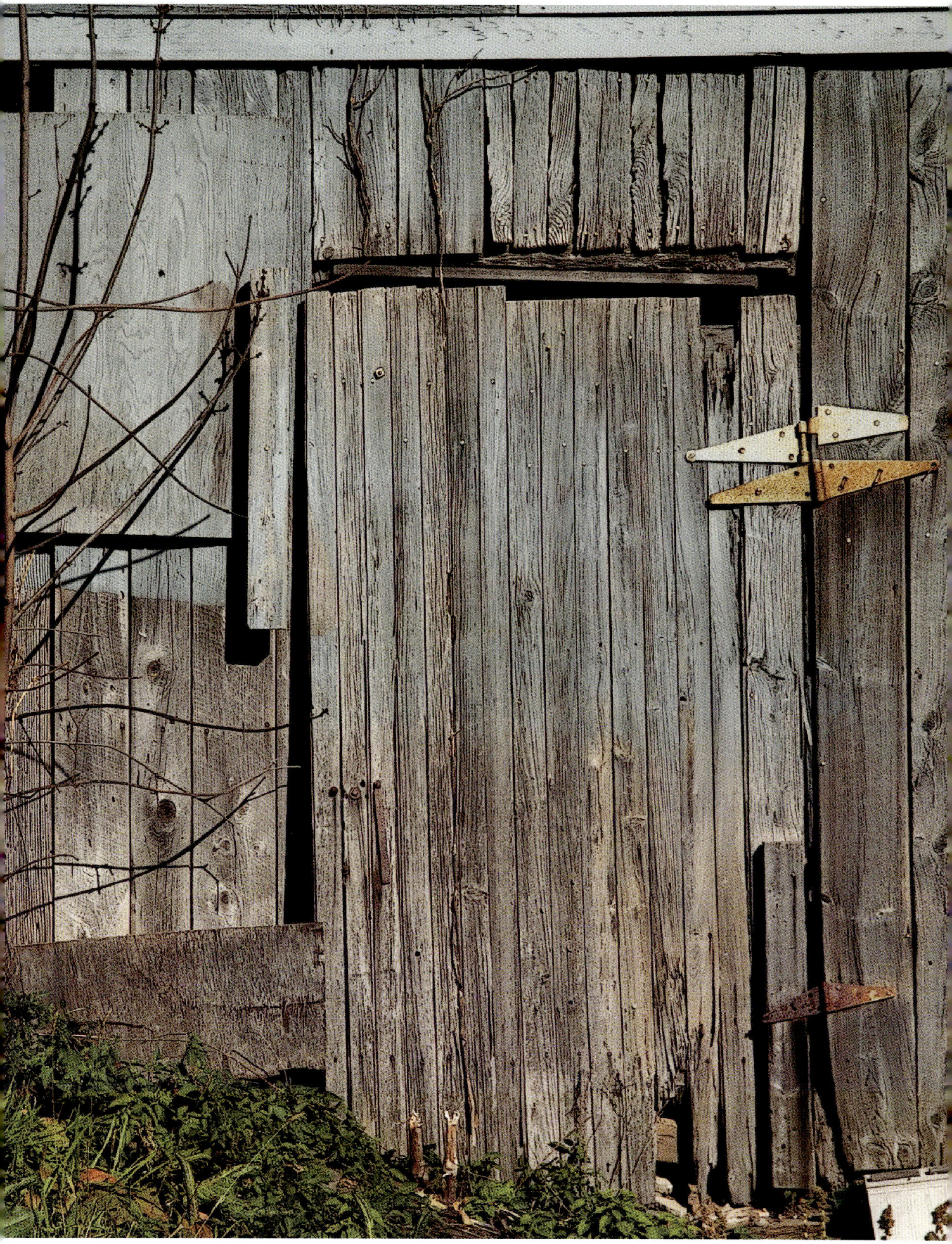

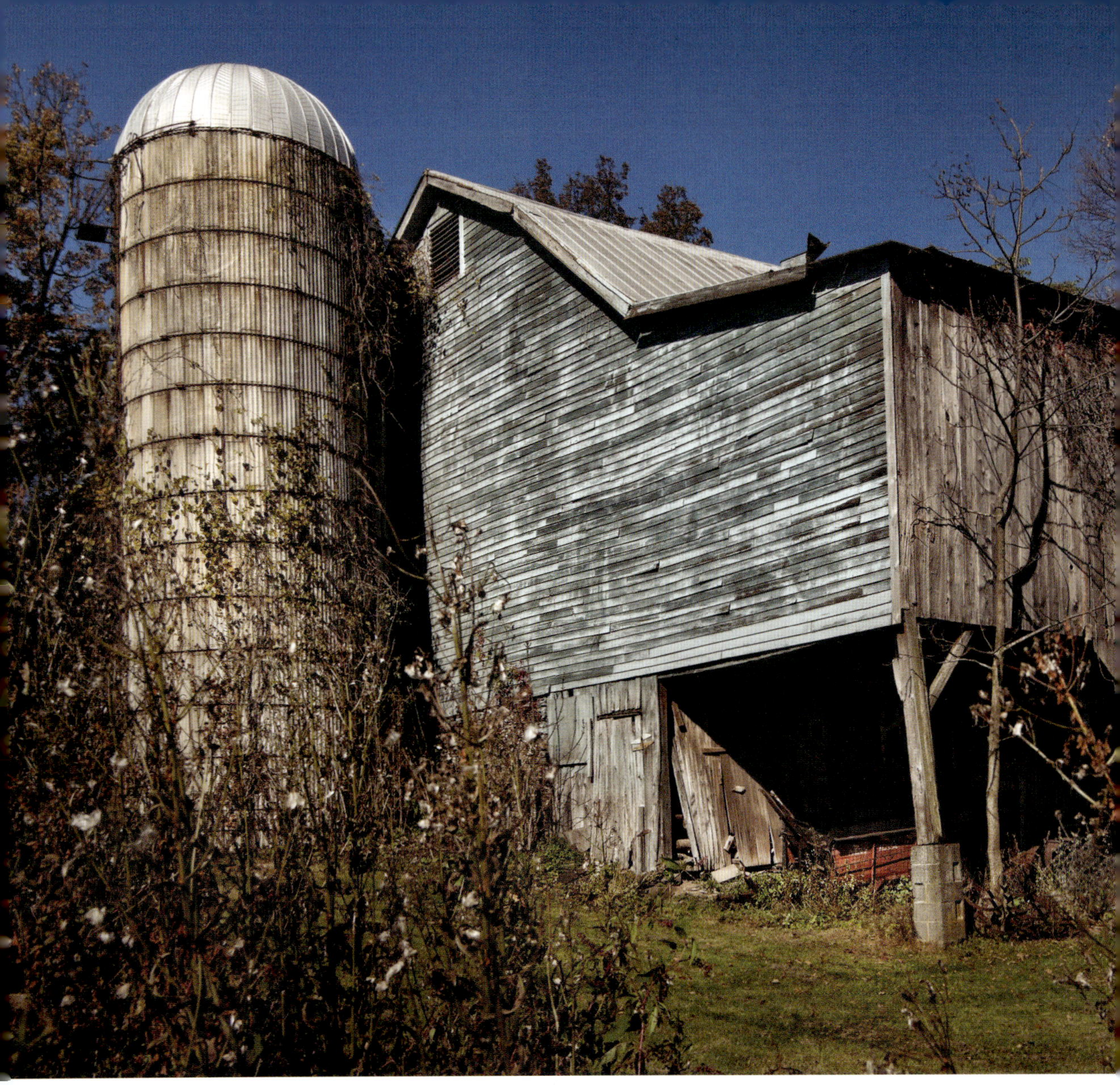

## Barn Colors

Linseed oil derived from flax plants was used to seal wood against rotting. The oil stained the wood a dark blue. Farmers noticed that painting their barns with the homemade paint kept the buildings warmer during the wintertime.

As mentioned previously, traditional tobacco barns in Kentucky were painted black to raise the heat inside the barn, aiding in the curing of the tobacco.

To achieve the iconic barn red color, farmers combined rust, milk, lime, and linseed oil. They then painted the barns this red hue to seal and preserve the barn wood.

Still other barns were painted white.

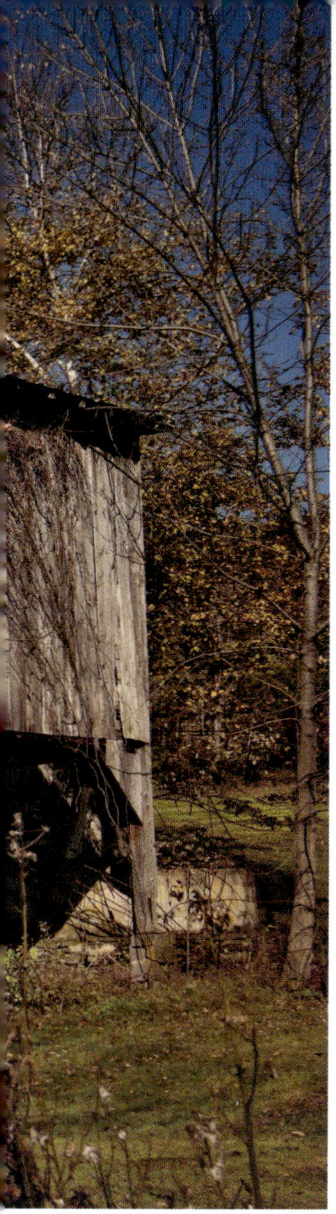

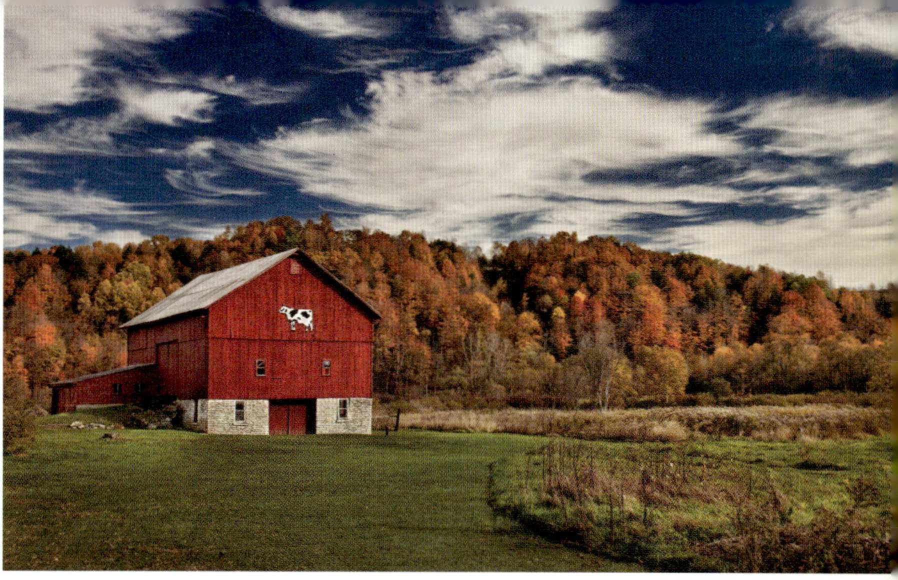

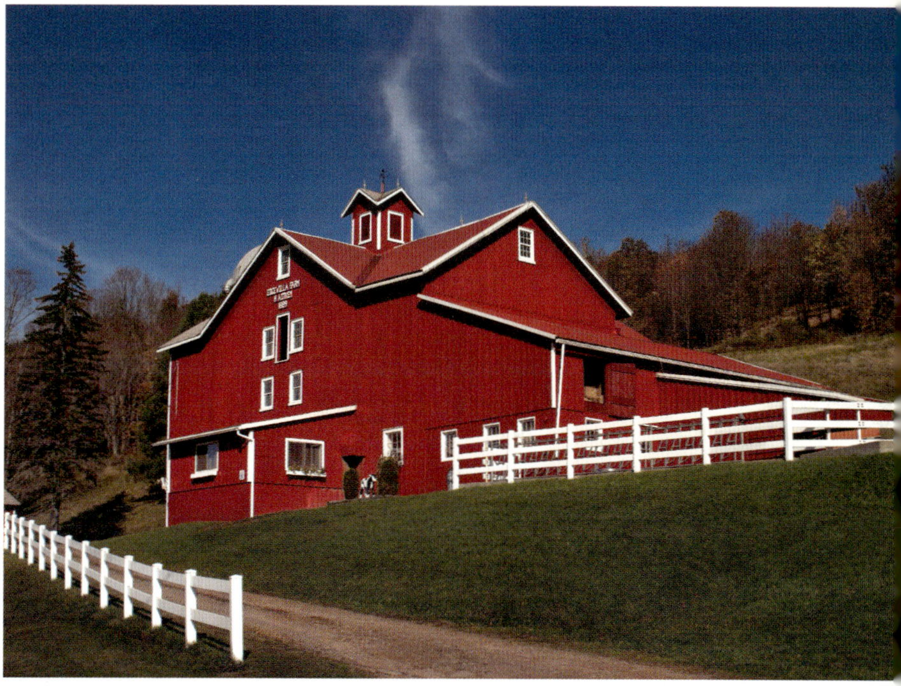

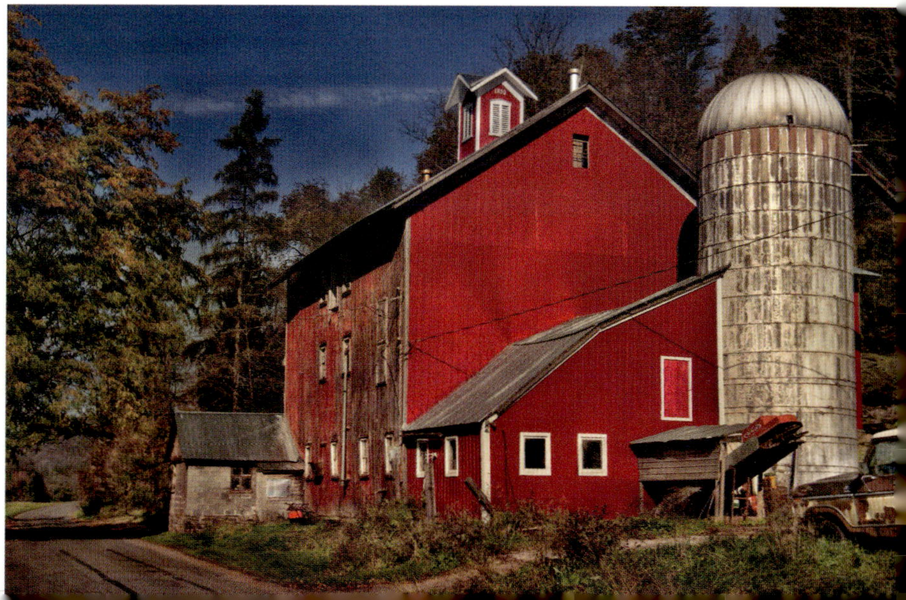

**previous page**—Blue barn. Pennsylvania.

**this page**—Three classic red barns. New York.

**following page spread**—A well-maintained red barn in Pennsylvania.

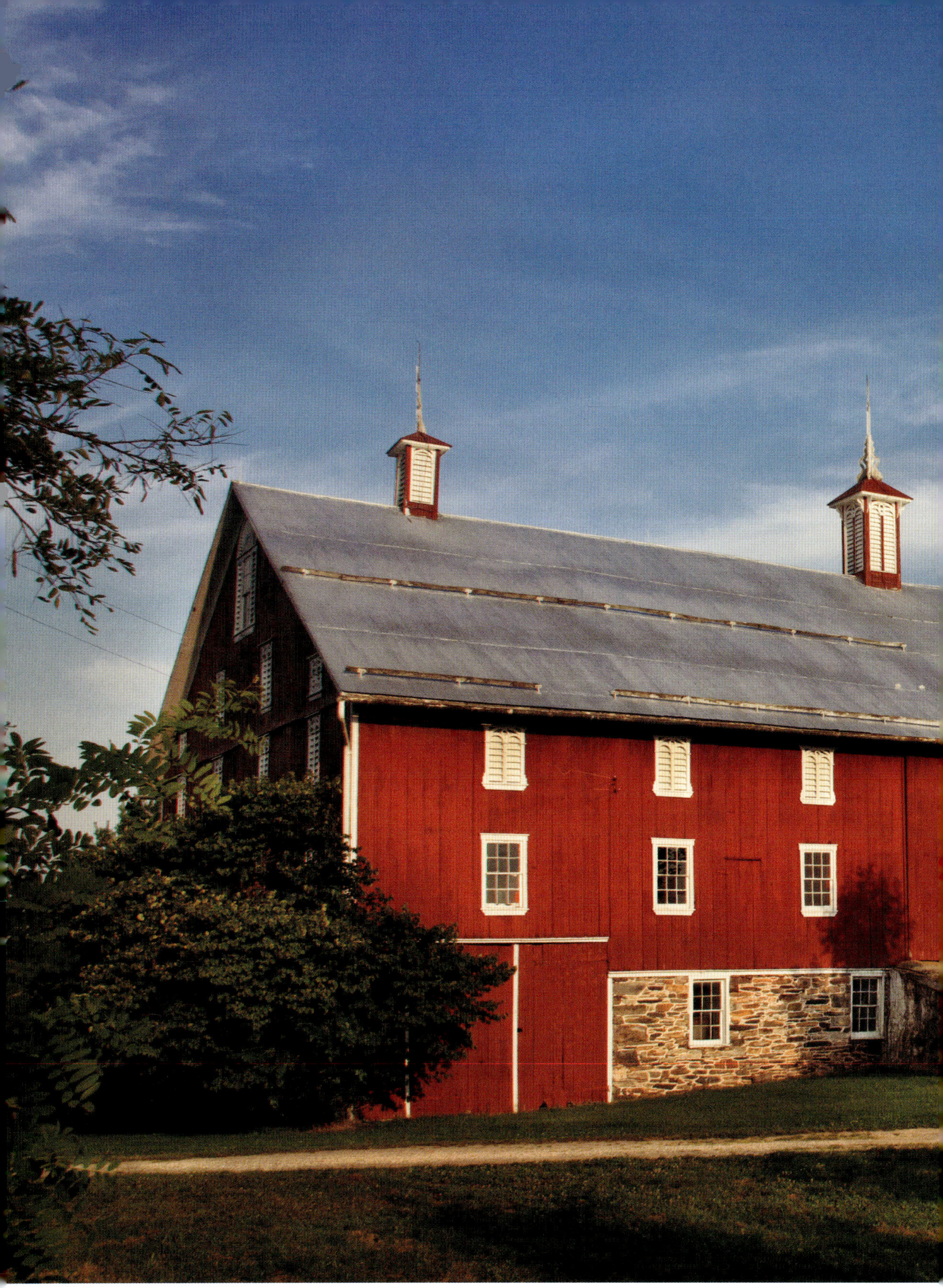

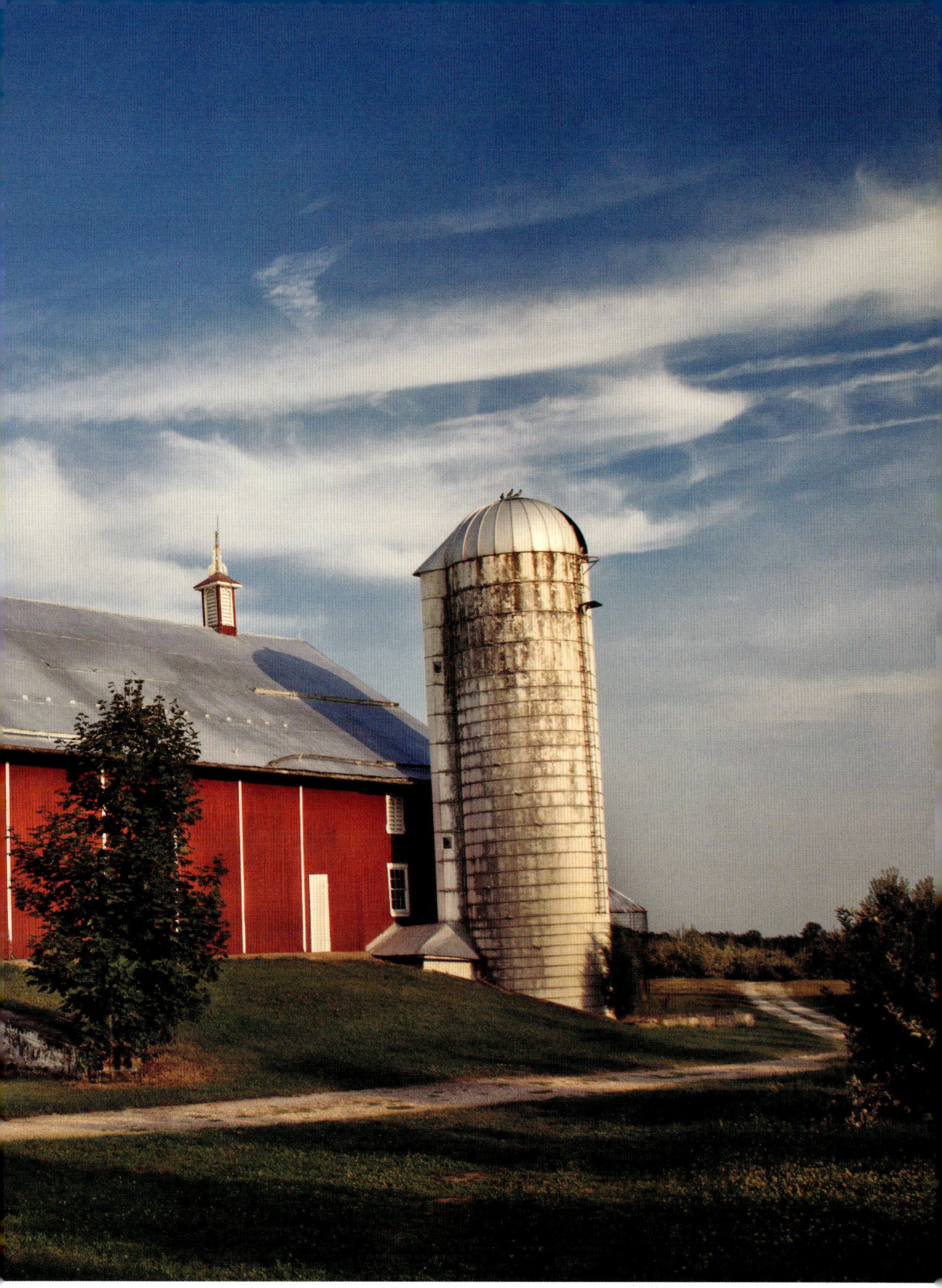

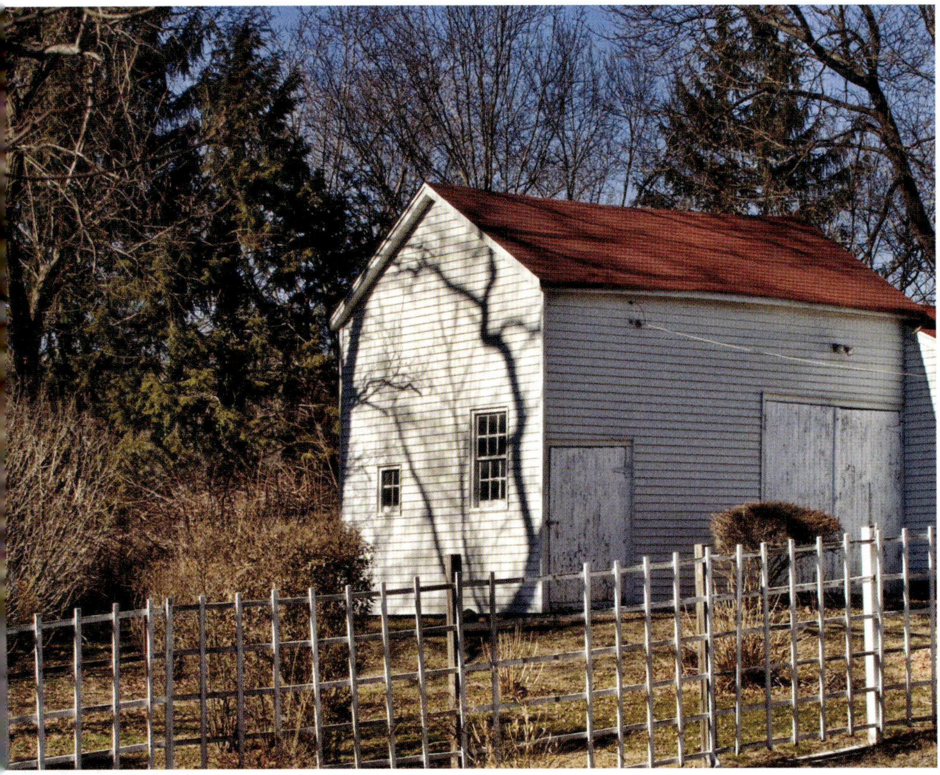

**left**—Ridgewood, New Jersey.

**below**—Pennsylvania.

**following page, top**—Colorado.

**following page, bottom**—Pennsylvania.

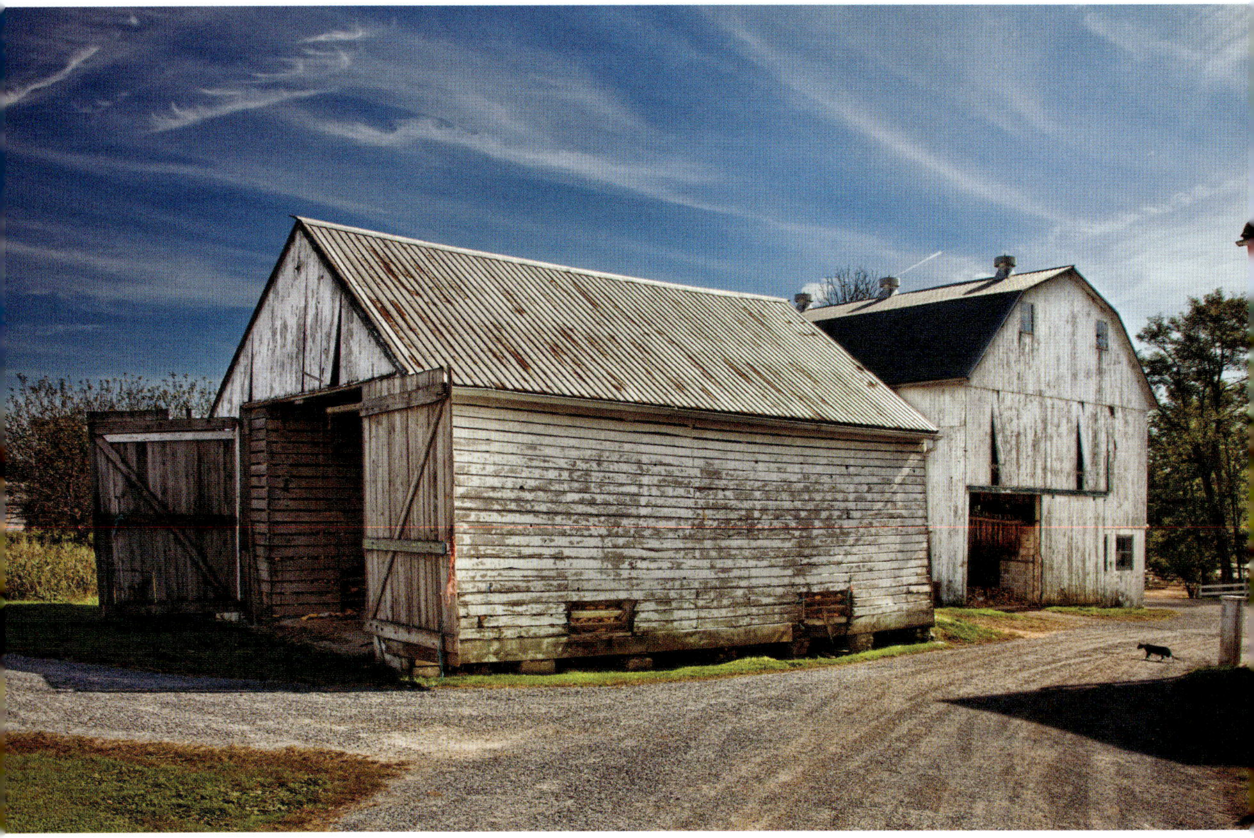

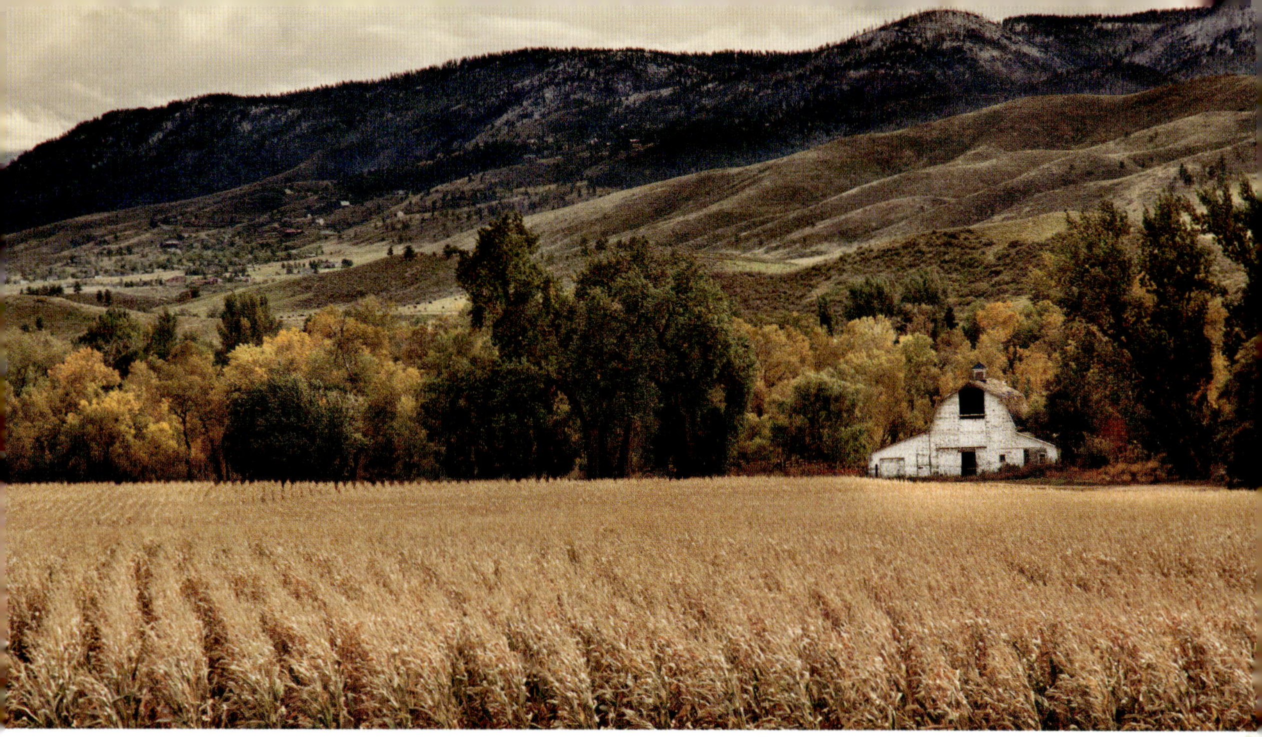

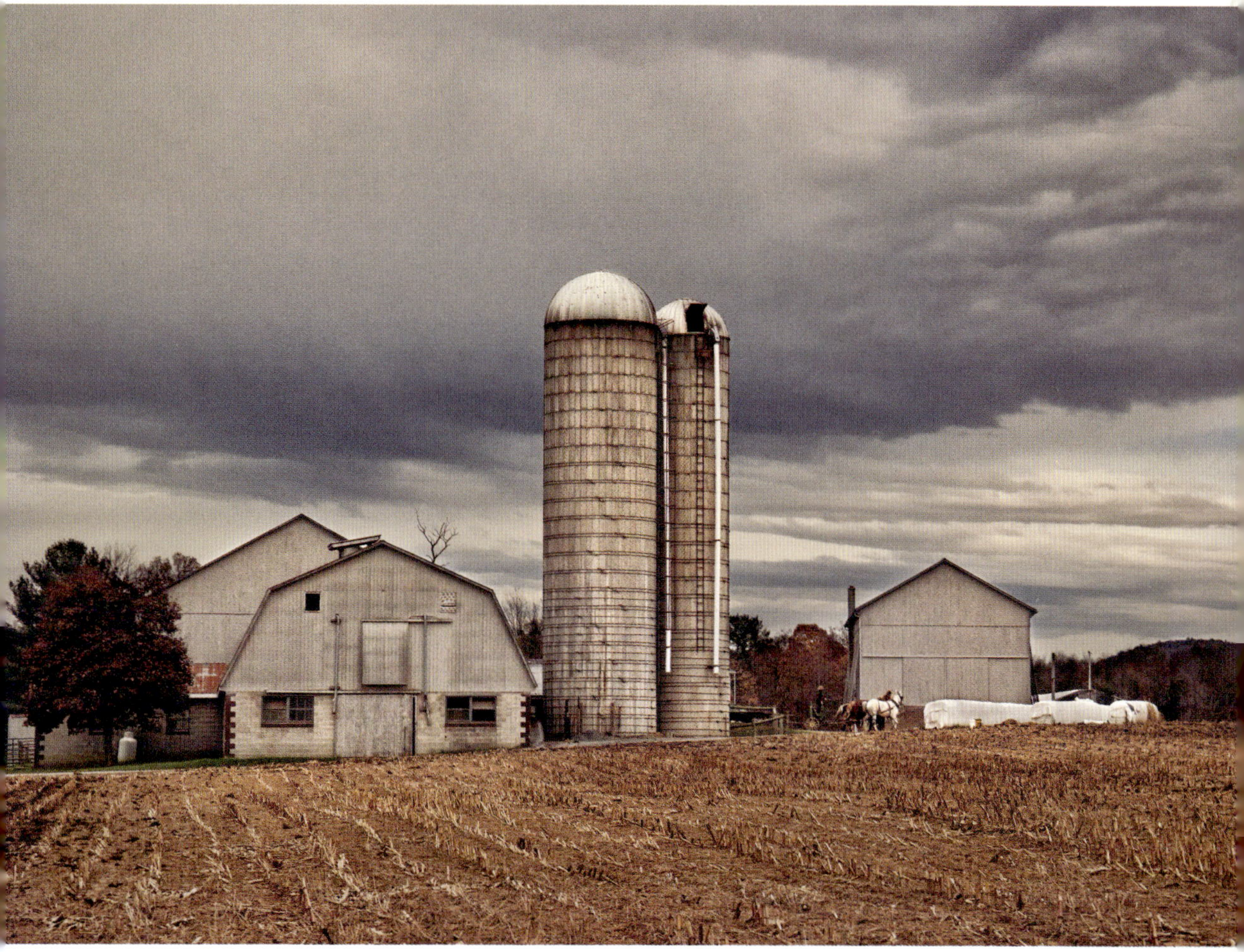

# A Personal Touch

Across America, proud farmers often added a personal touch to their barns and farms. In this chapter, you'll find images that show how owners infused their properties with a little bit of personality.

## Hex Signs, Barn Quilts, and Art

Pennsylvania Amish/German farmers put colorful symbols called Hex signs on their barns to keep the evil spirits away. "Hex" means "witch" in German.

Do not confuse hex signs with barn quilts. Barn quilts are a large piece of wood that is painted with colorful patterns to mimic a quilt. They are then hung on the side of the barn as art.

**below**—Barn quilts. Kentucky. **following page, top and bottom**—Barn quilts.

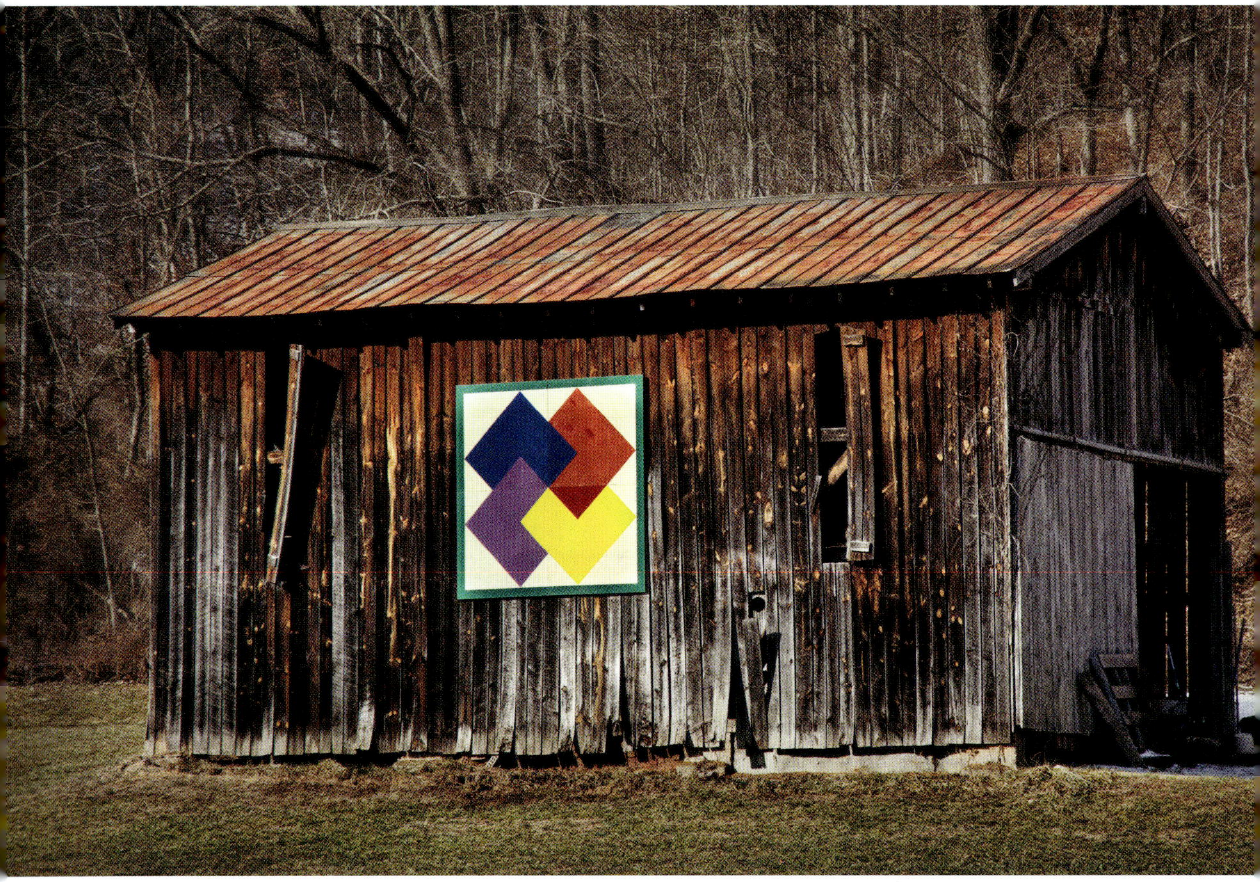

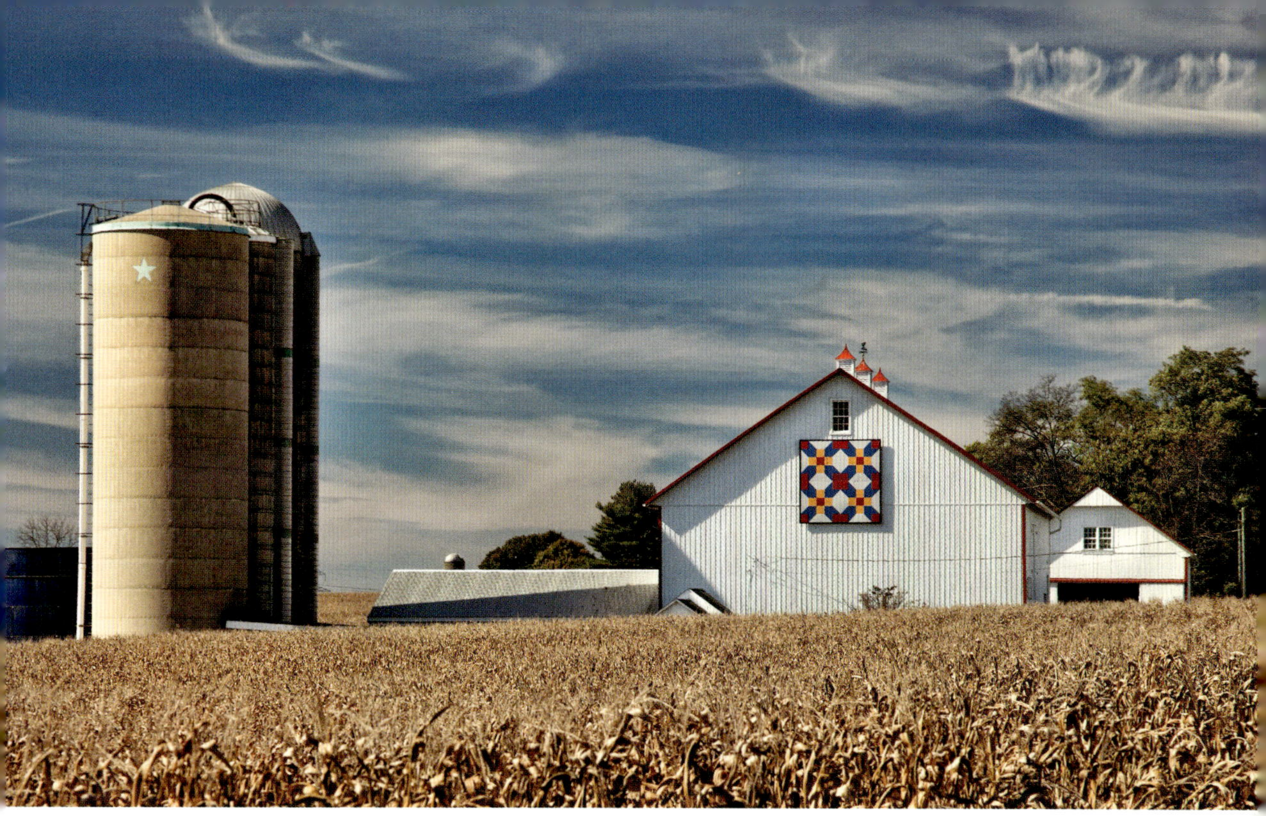

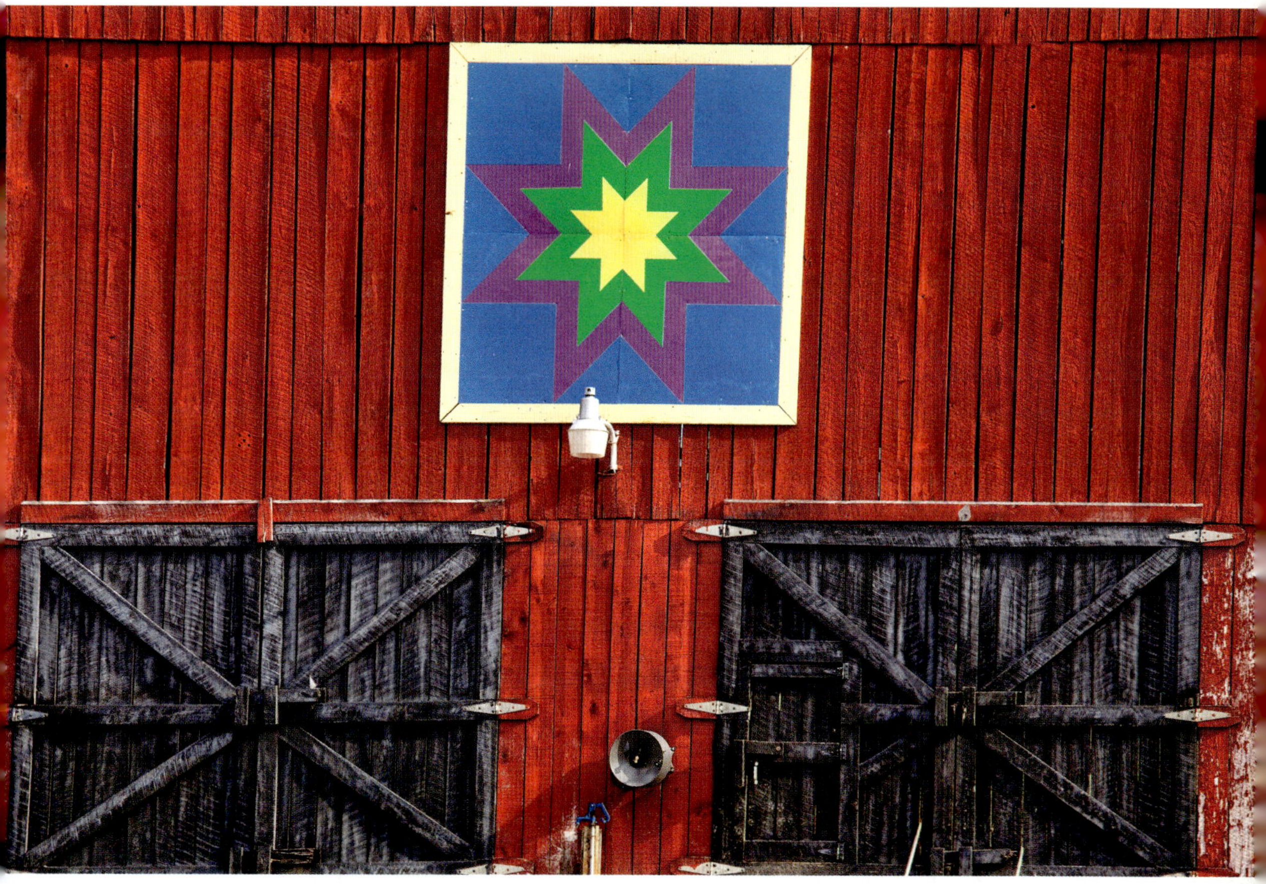

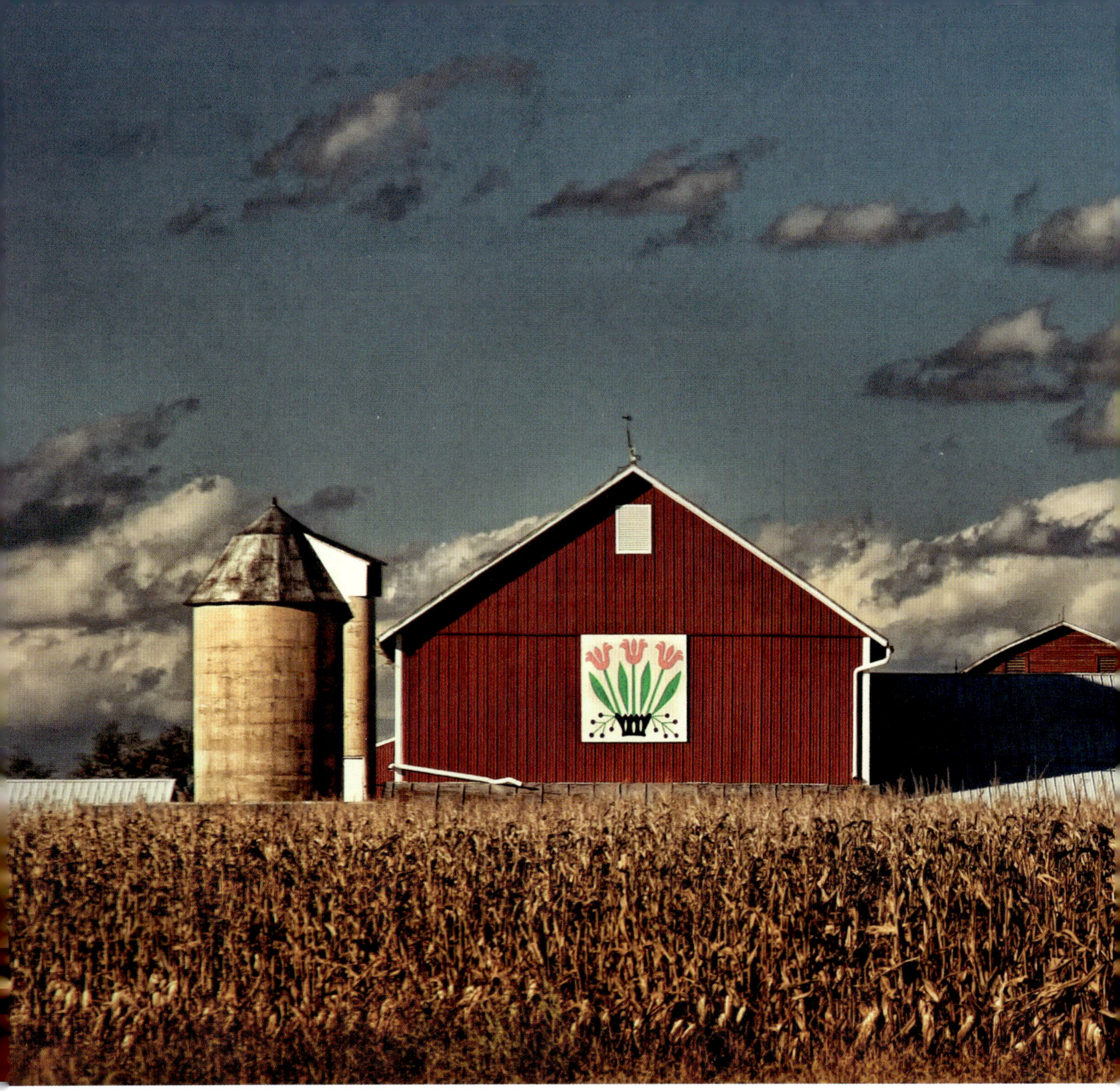

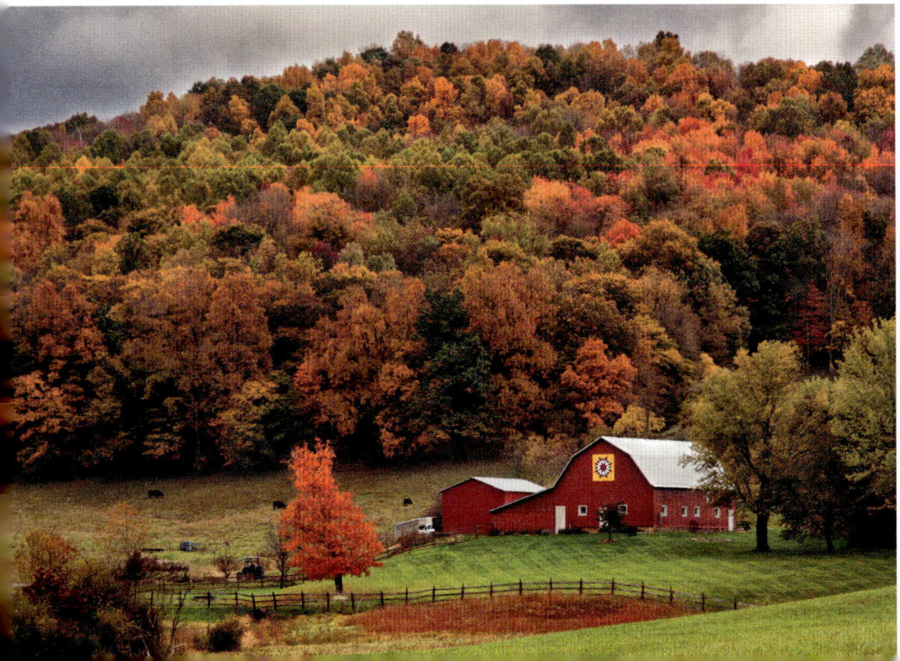

**above**—Barn quilt. Pennsylvania.

**left**—Barn quilt. West Virginia.

**following page, bottom**—Barn quilt. Vermont.

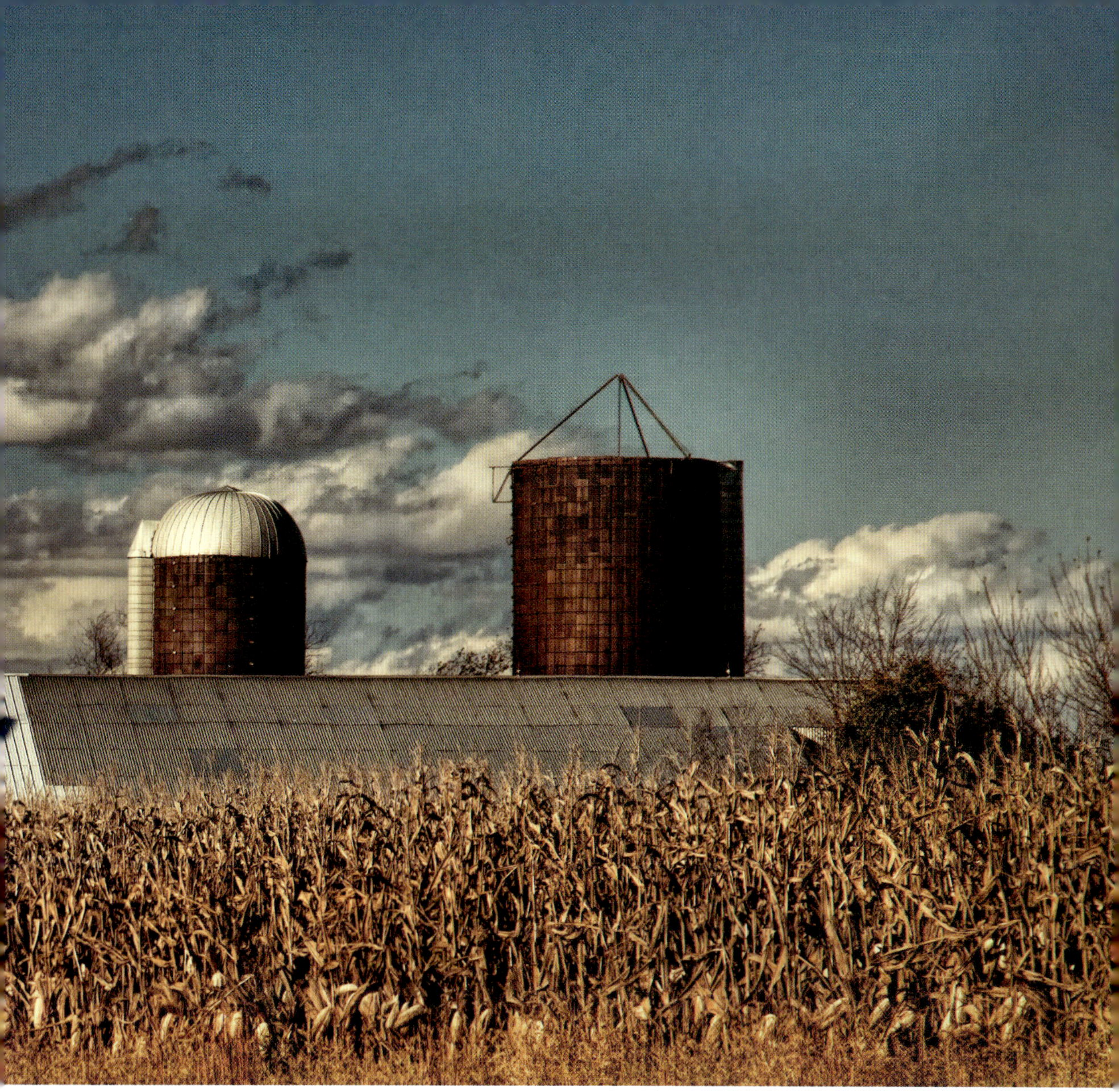

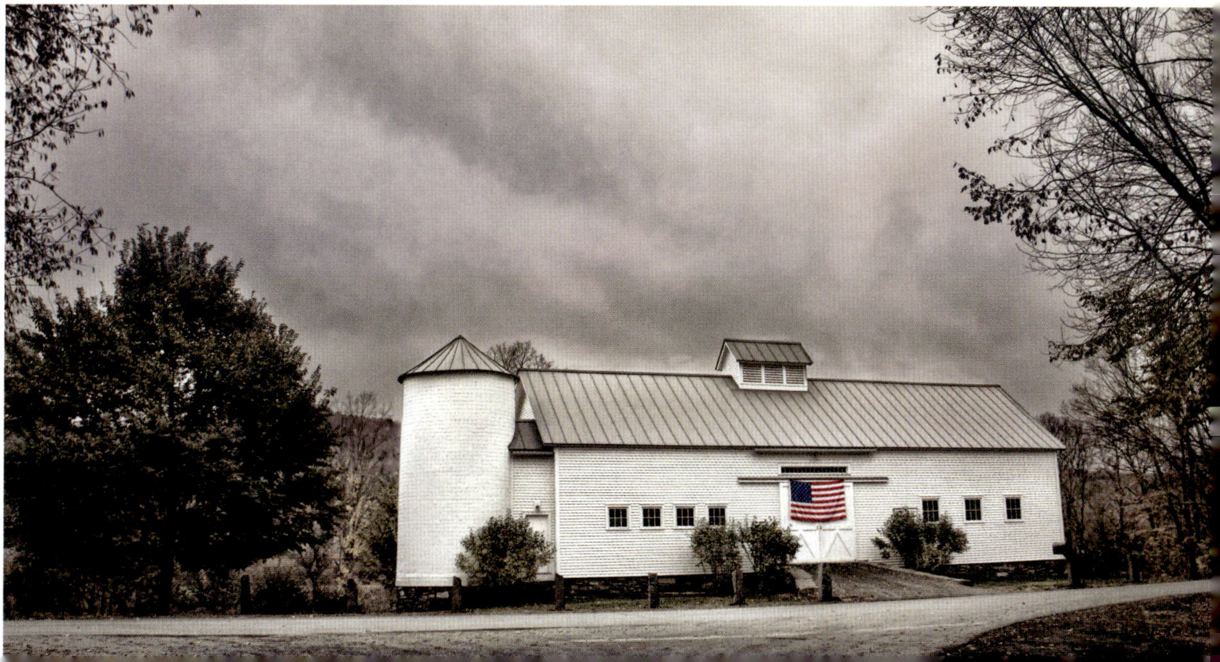

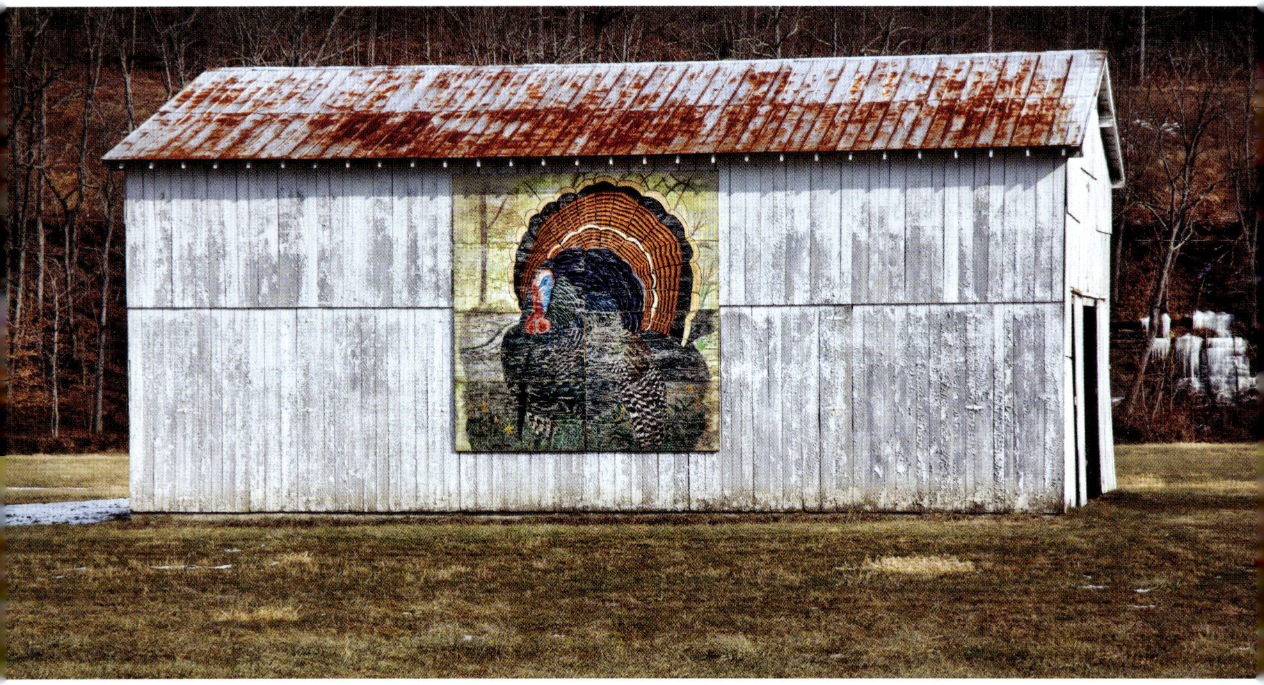

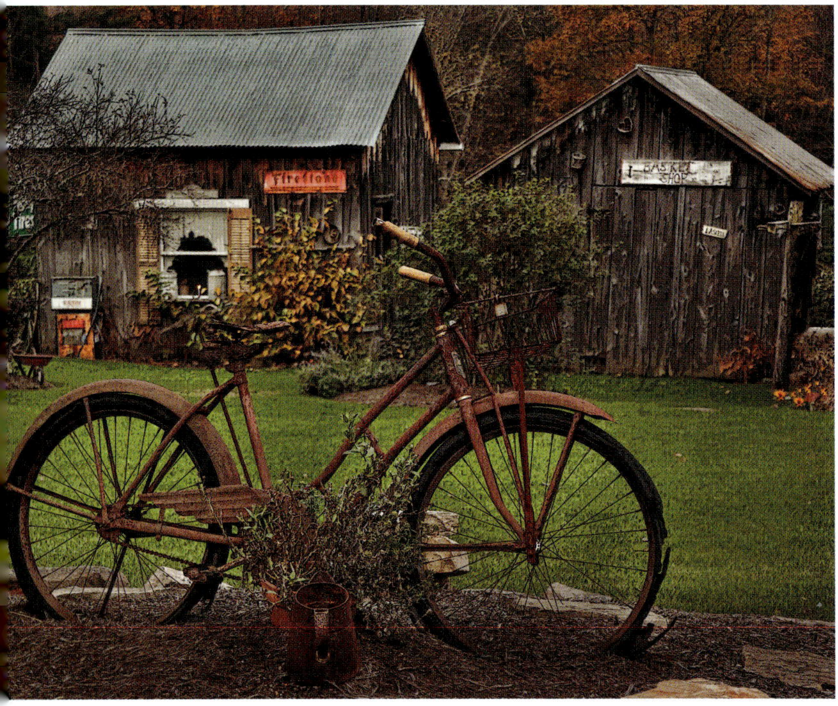

**top**—Kentucky.

**bottom left**—Barn art. West Virginia.

**bottom right**—Vermont.

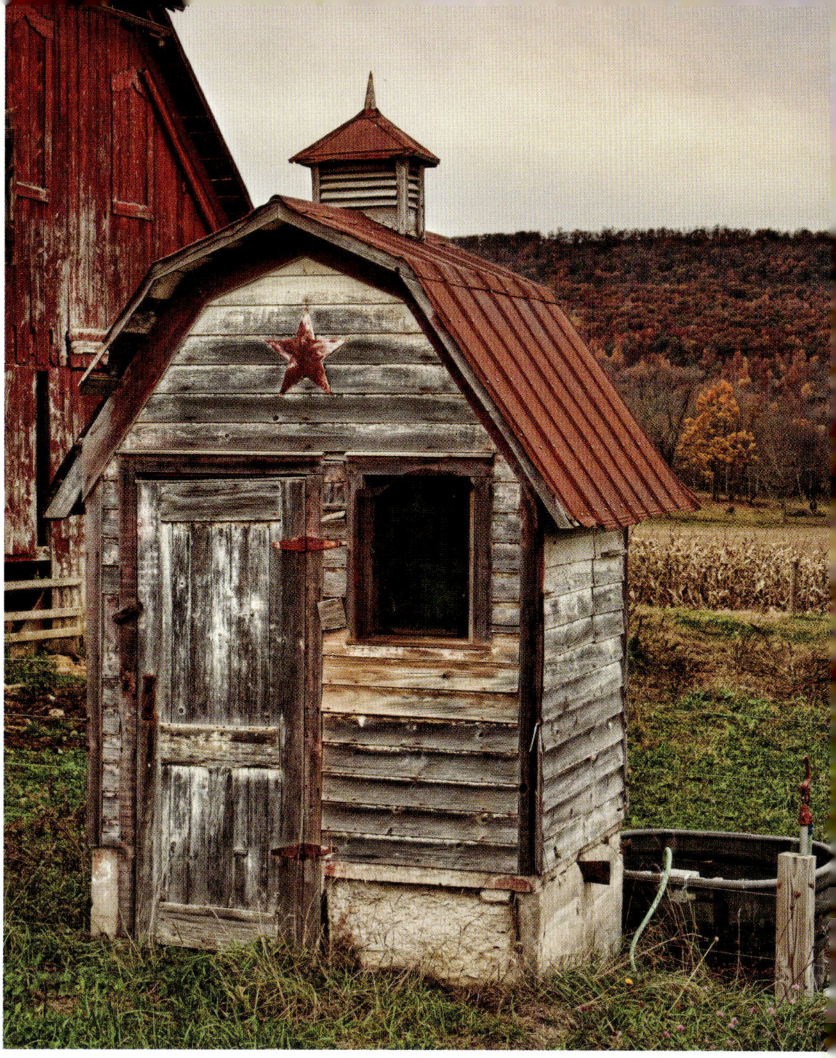

**right**—Pennsylvania.

**below**—Barn quilt. Virginia.

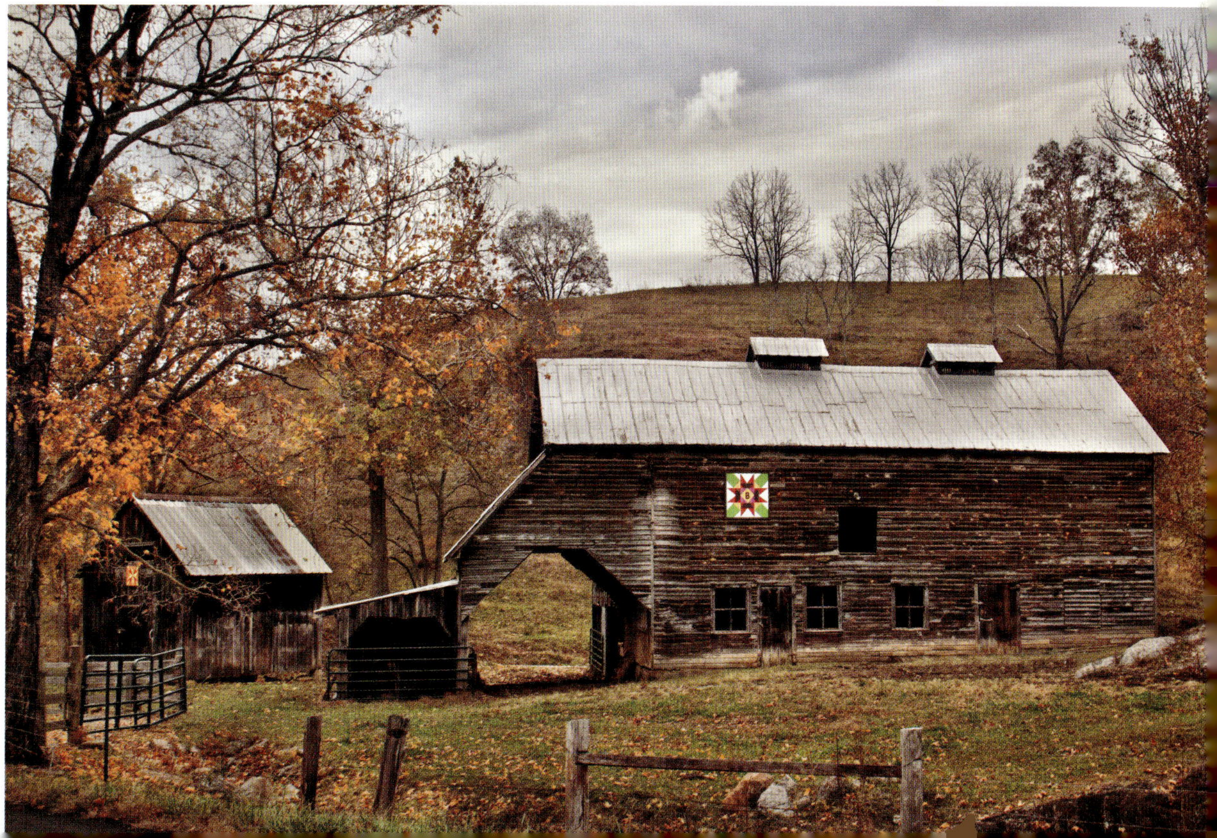

# Advertisements

The Bloch Brothers Tobacco Company is credited with starting barn advertising in 1890. By 1925, Mail Pouch barns started to develop. By the early 1960s, some 20,000 barns in 22 states displayed the Mail Pouch ad.

Sign painters took advantage of the size and visibility of barns in an age before billboards were used to raise product awareness.

As you travel across the country, you'll see a variety of barn advertisements for a wide range of products.

**below**—Wyoming.

**following page, top and center**—Three Mail Pouch advertisements on barns in West Virginia.

**following page, bottom**—Massachusetts.

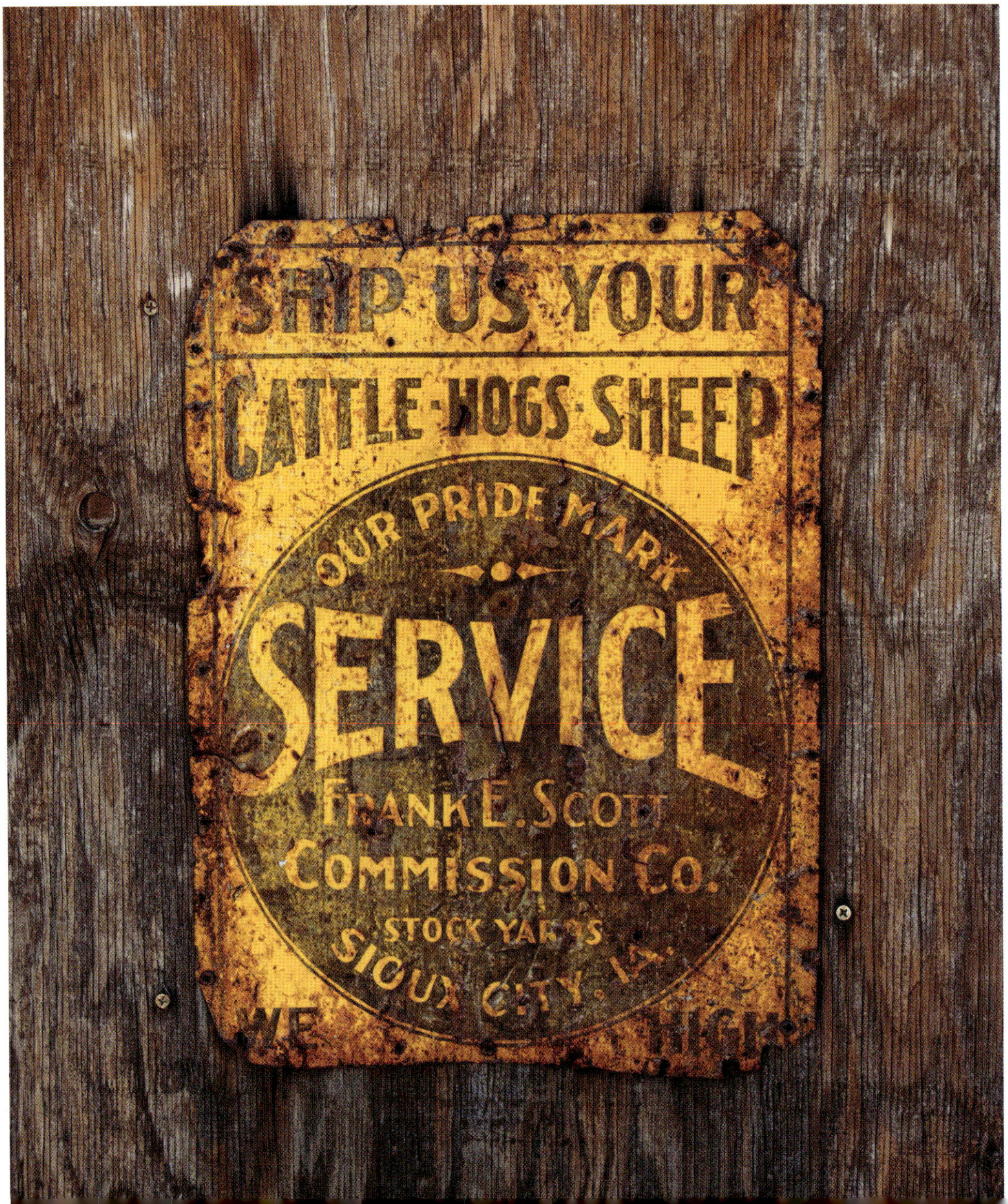

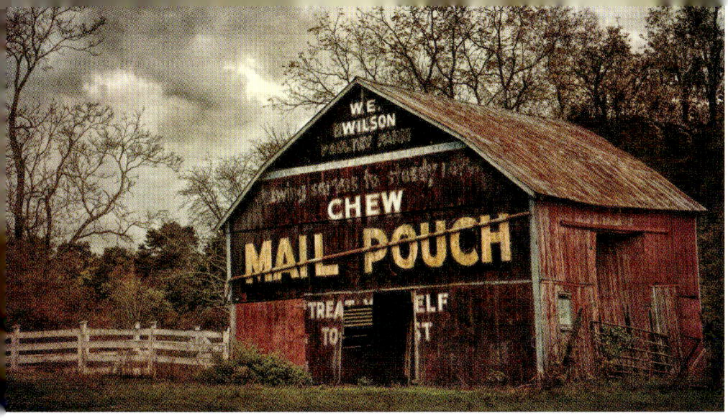
W.E. WILSON
CHEW
MAIL POUCH

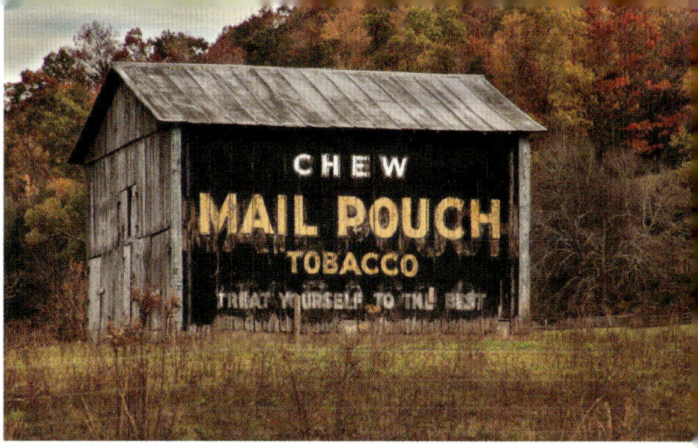
CHEW
MAIL POUCH
TOBACCO
TREAT YOURSELF TO THE BEST

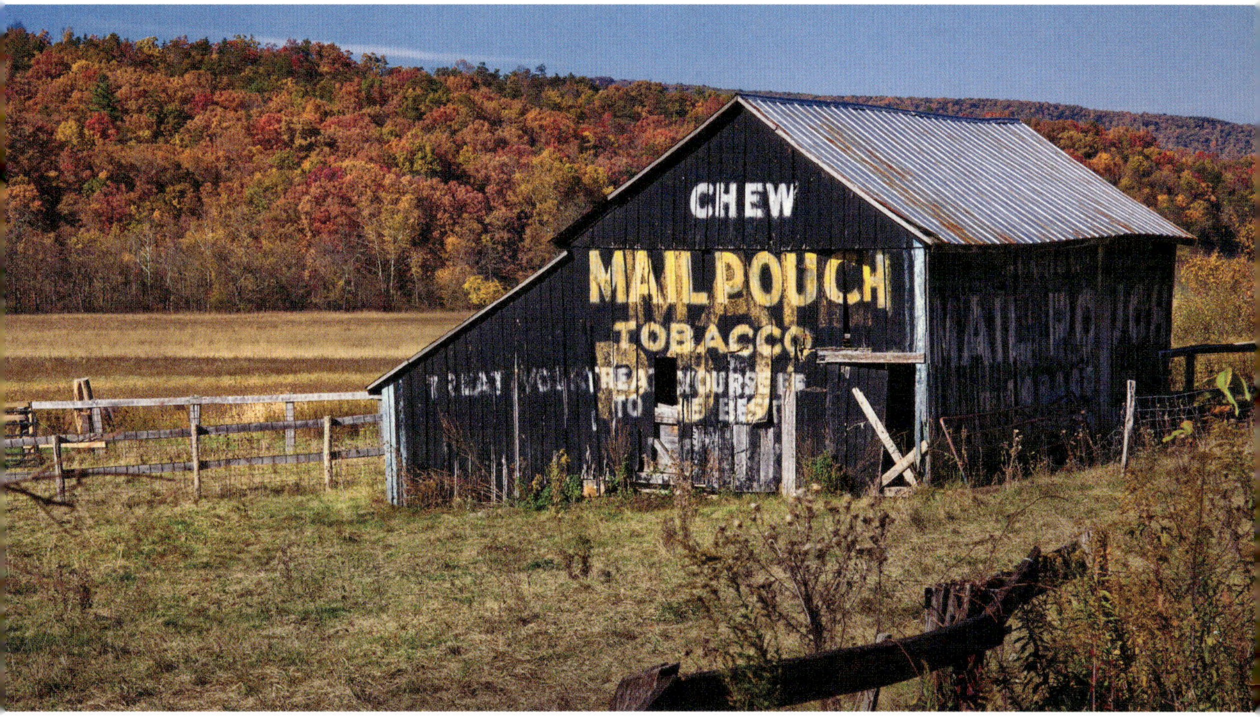
CHEW
MAILPOUCH
TOBACCO

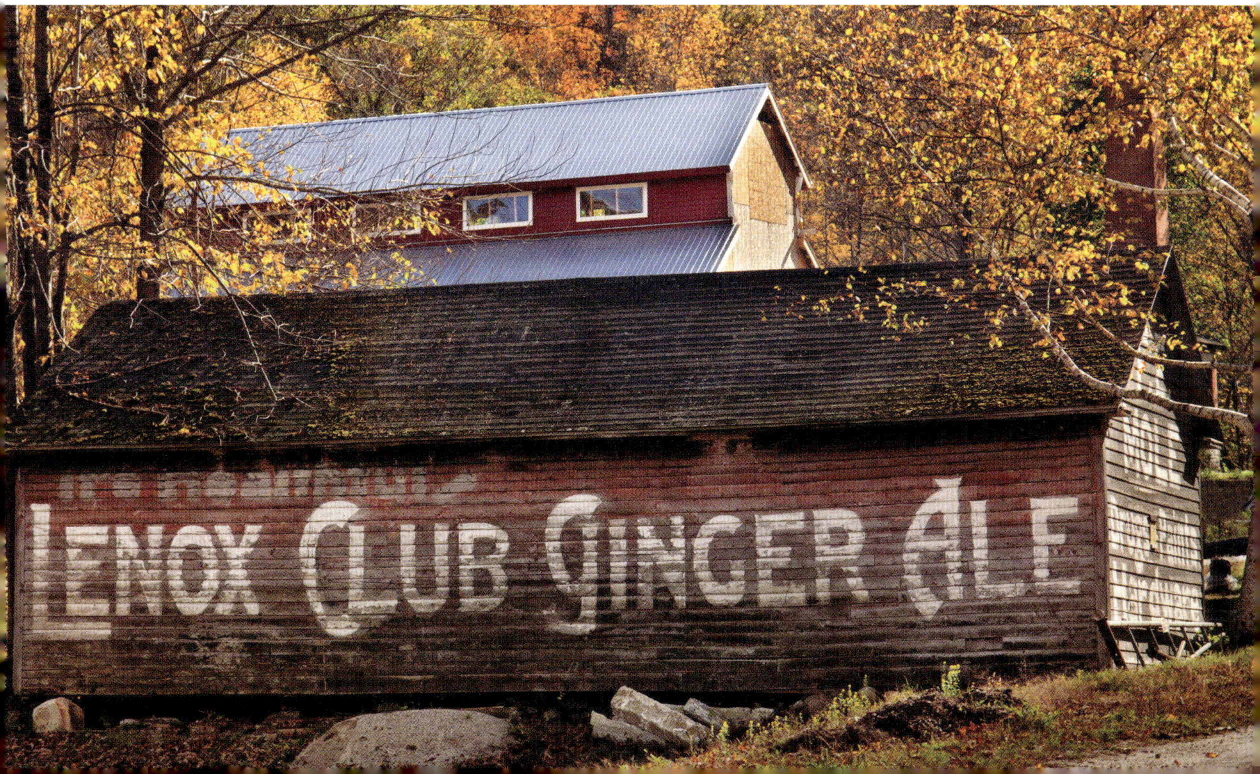
LENOX CLUB GINGER ALE

**this page and following page**—Advertisements on a smaller scale. Cordes Station, Arizona.

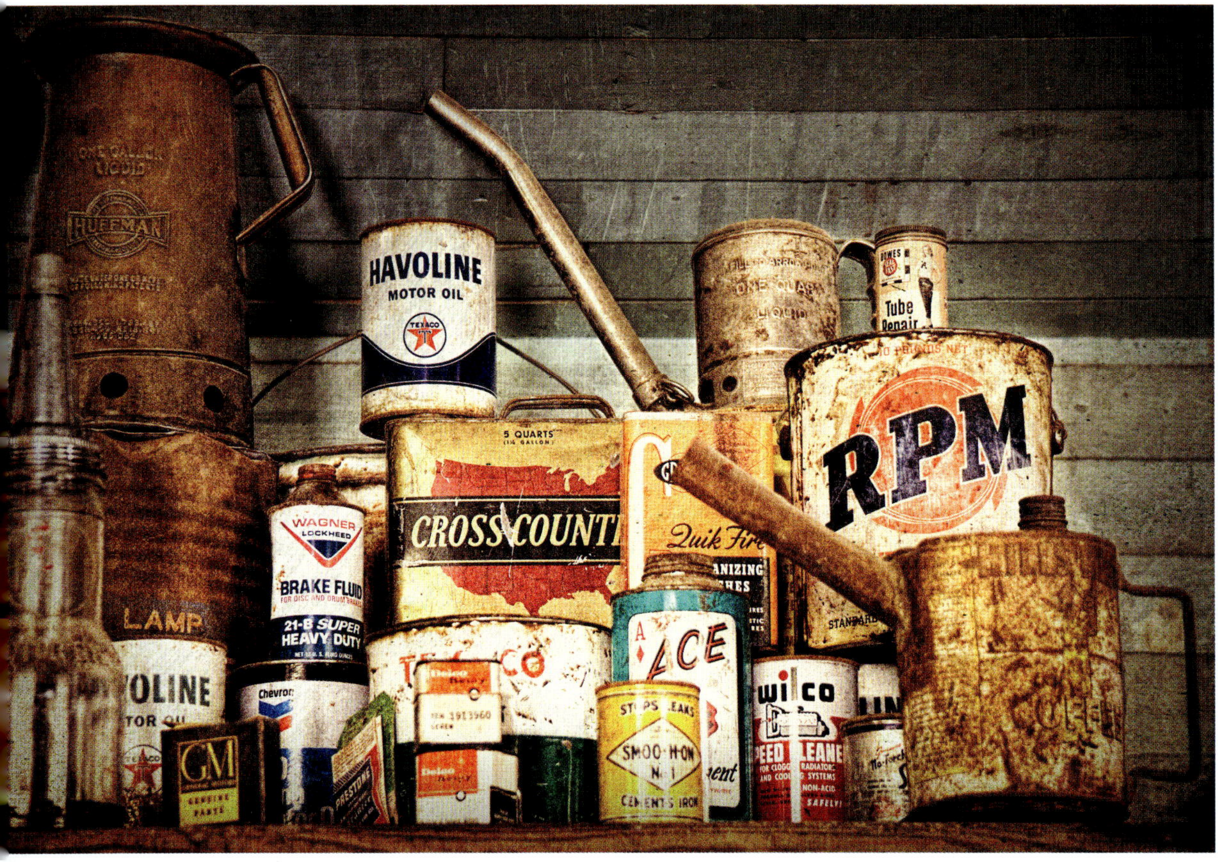


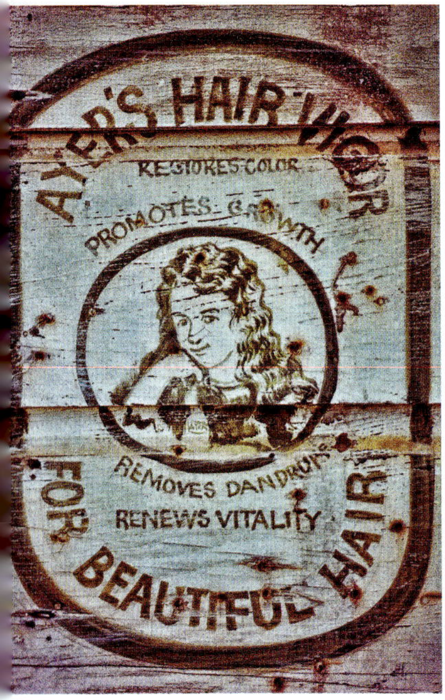


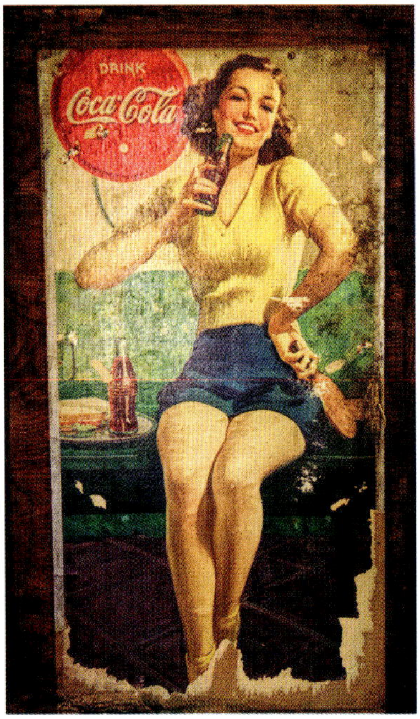


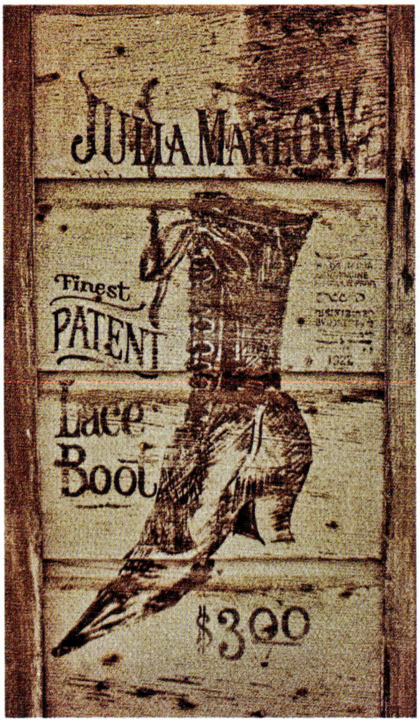

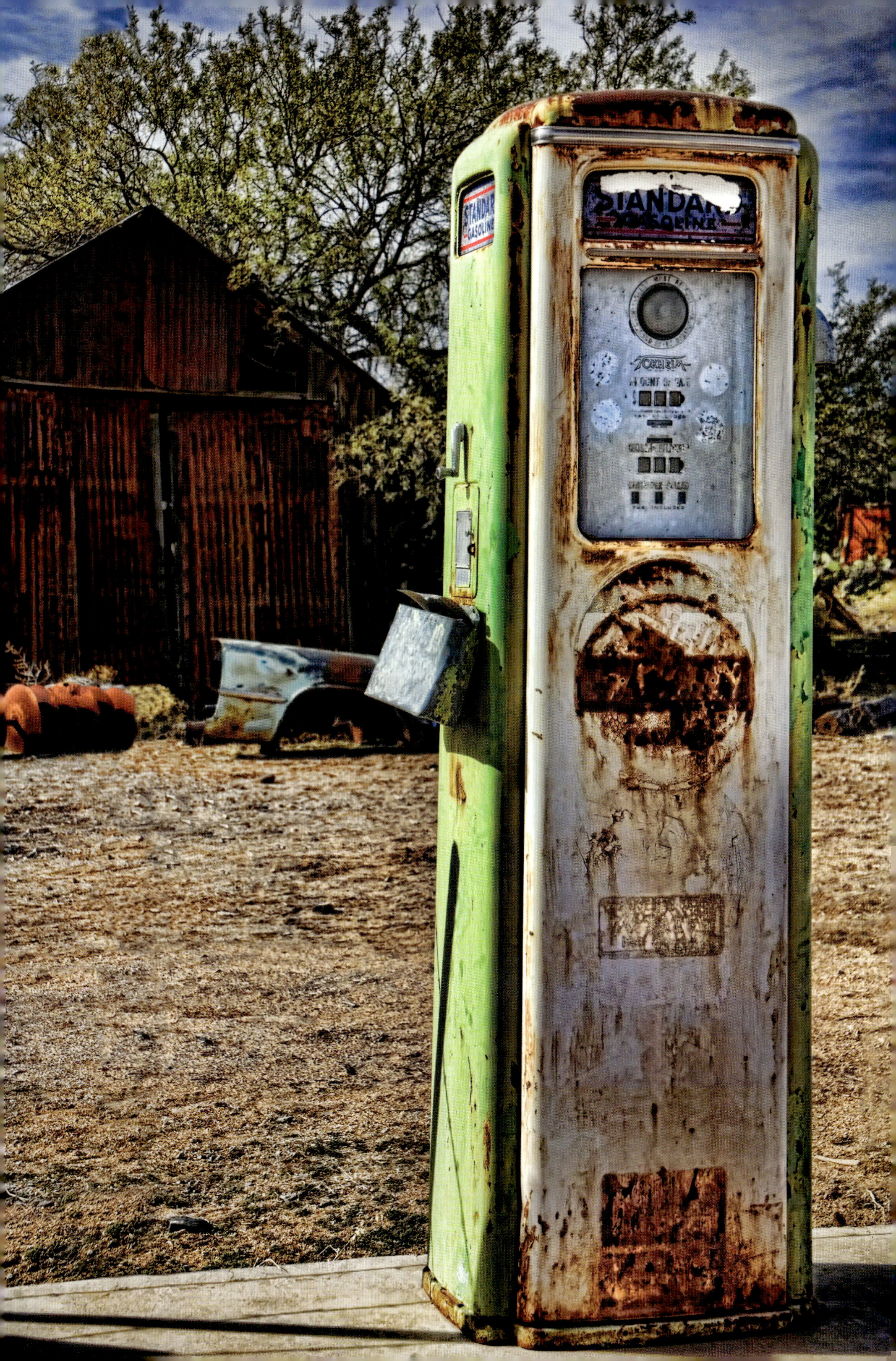

# Tools of the Trade

The weathered and rusty farm implements used by farmers years ago make for great photographic subjects. Every picture tells a story. Looking at the images, it's as if history comes alive.

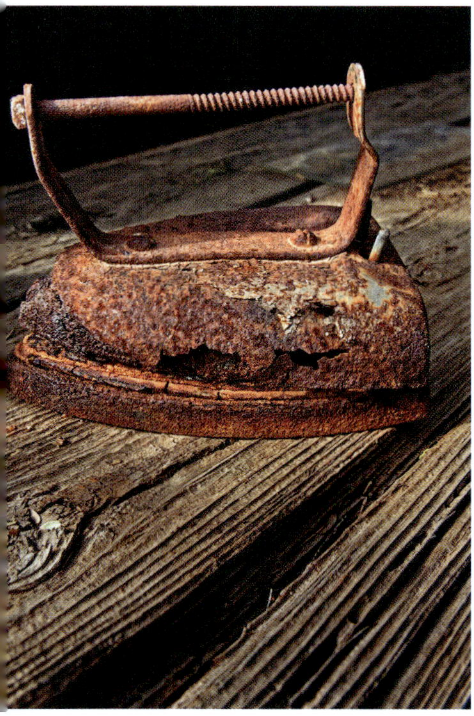

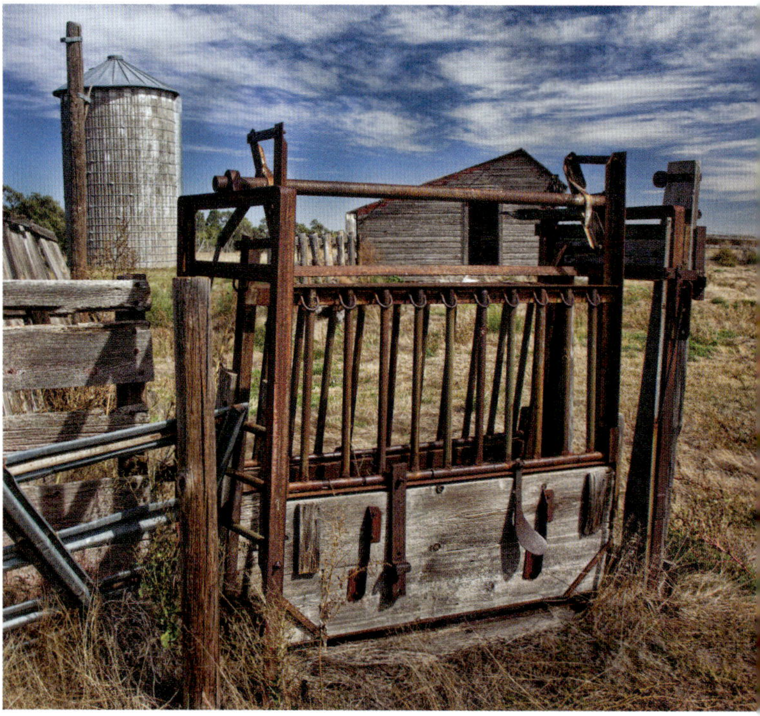

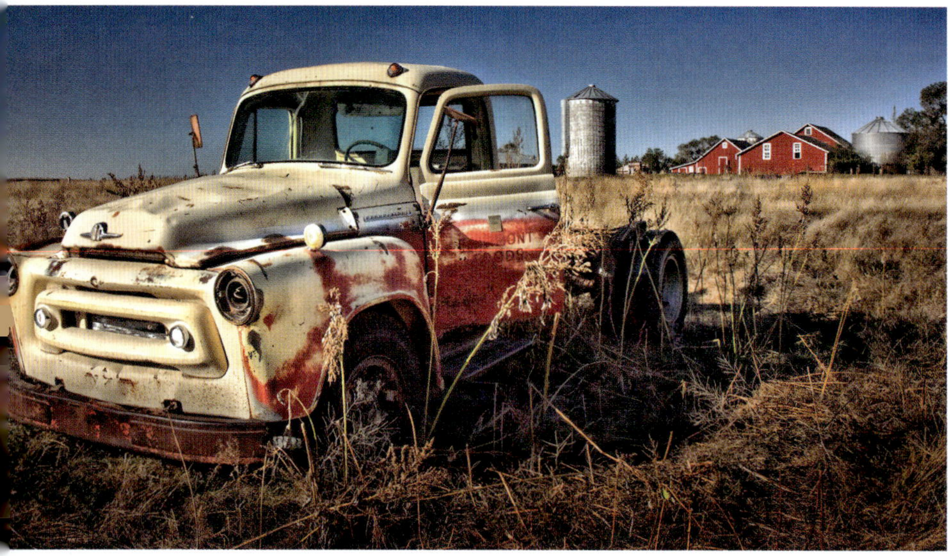

**top left**—Arizona.

**top right and bottom**—Nebraska.

**following page**—West Virginia.

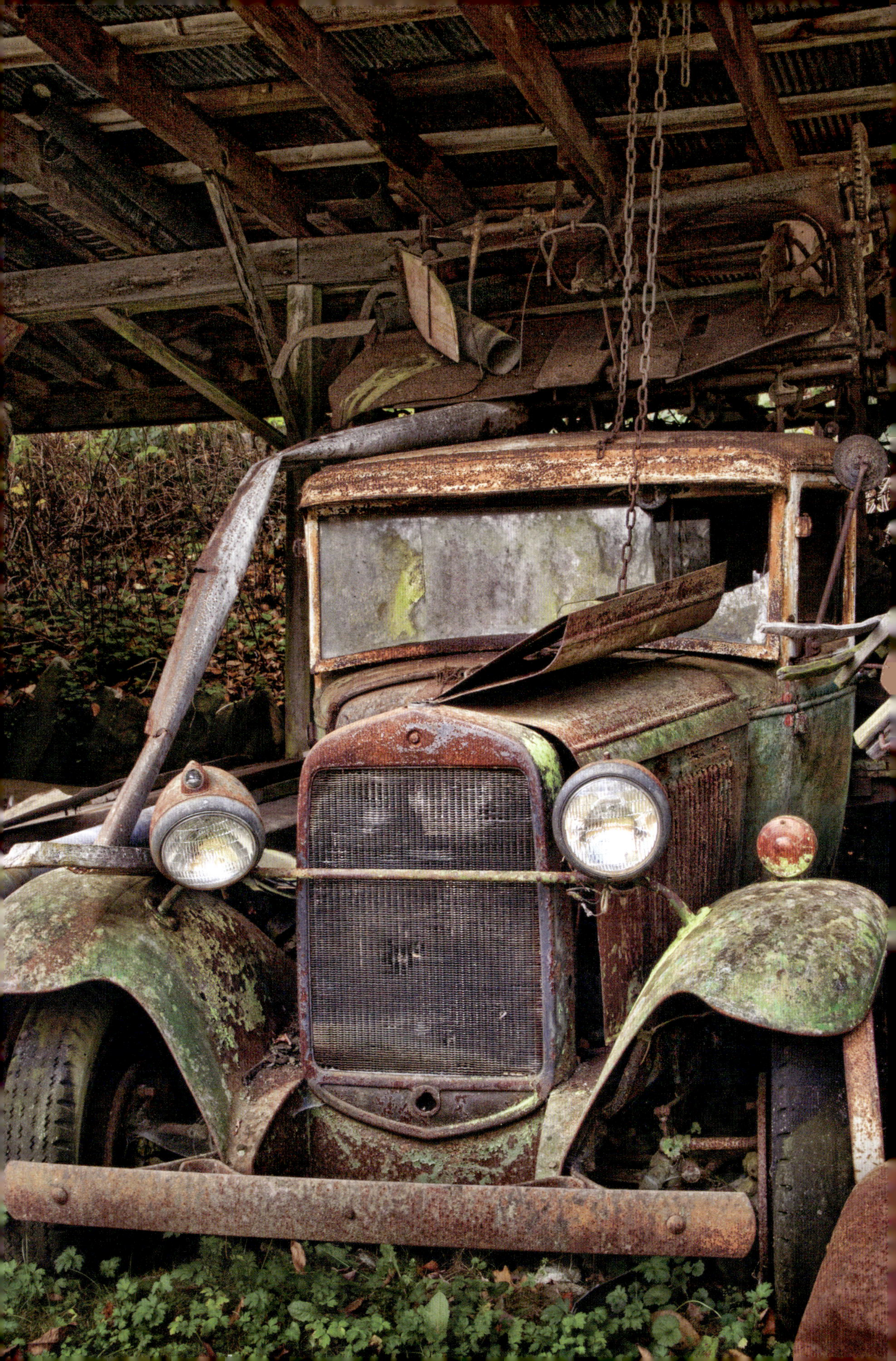

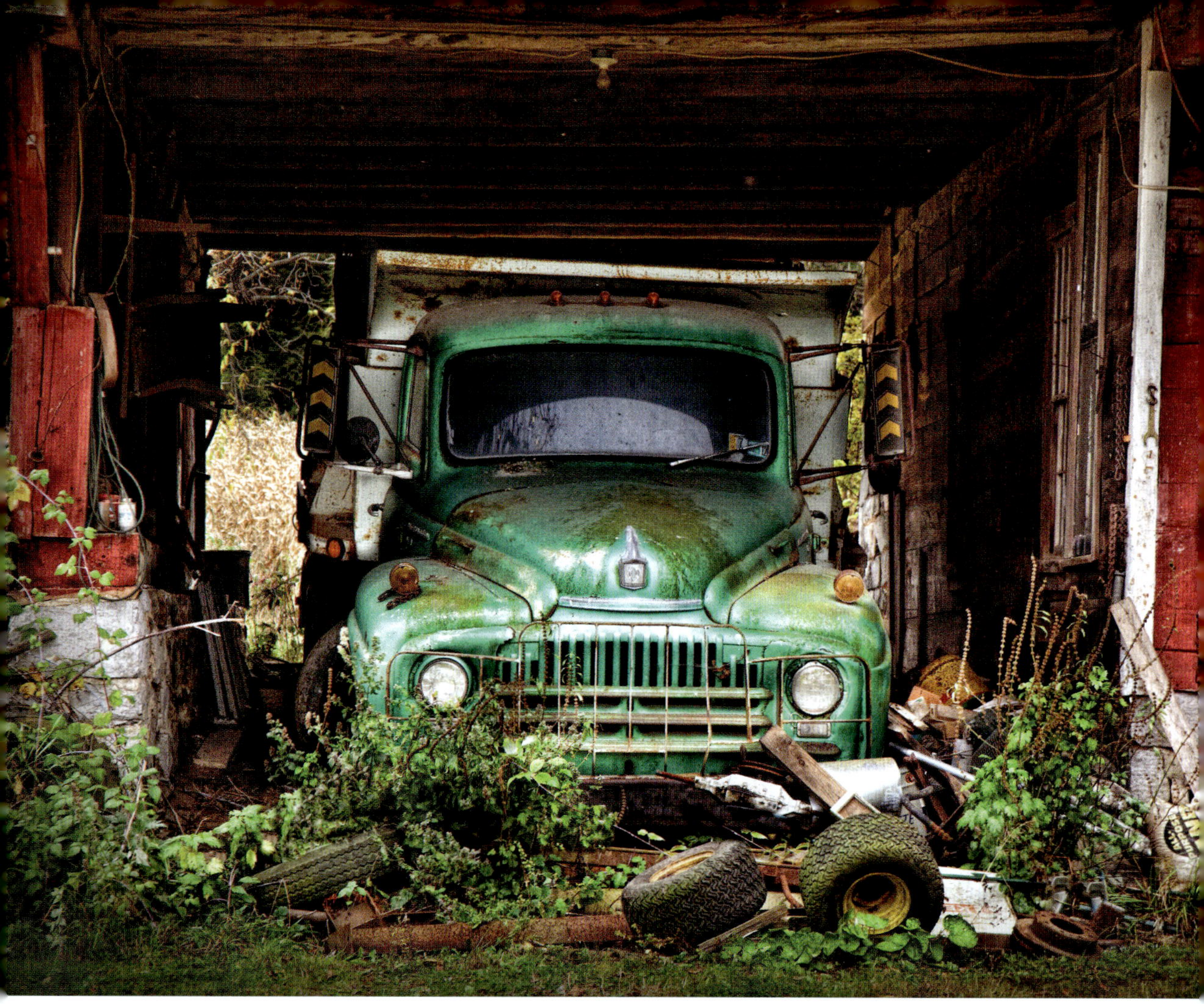

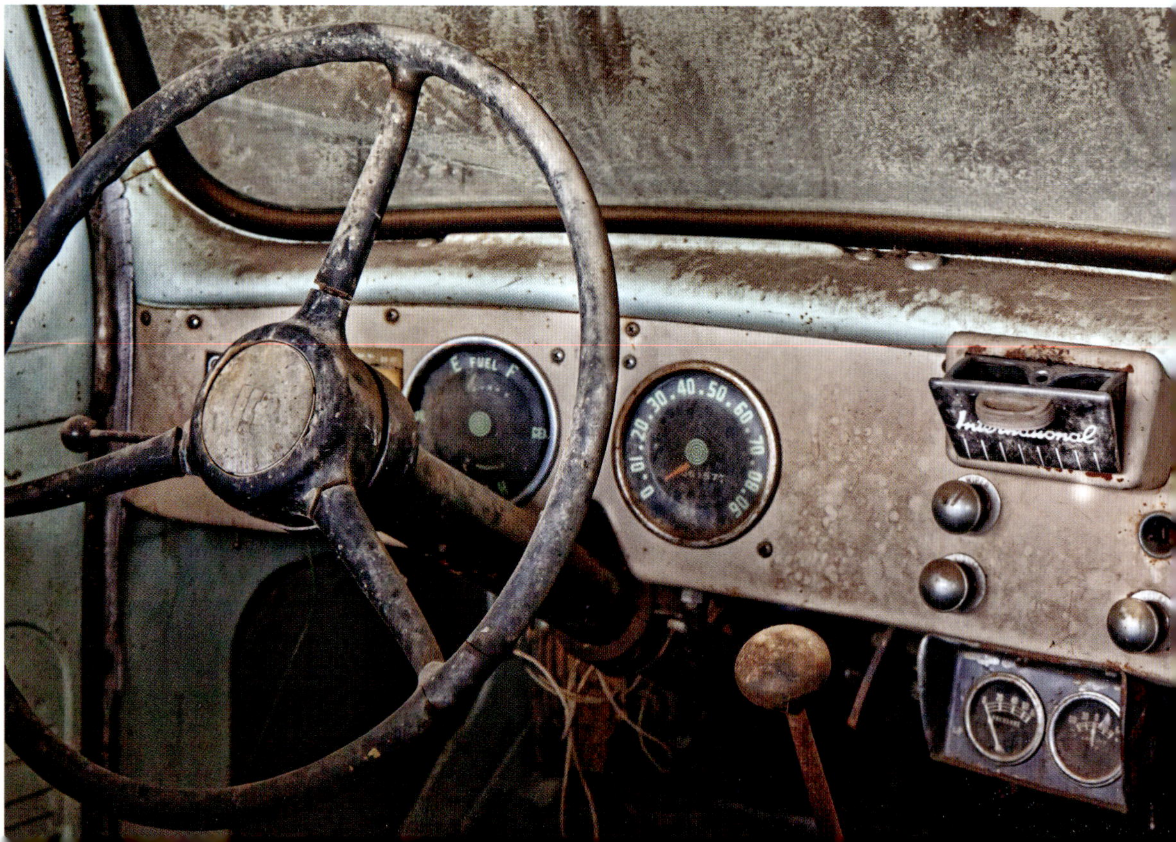
E FUEL F
International

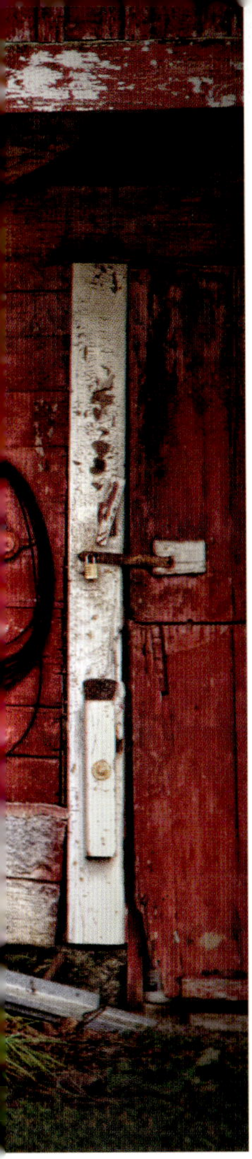

**previous page and right—** There's no denying that these vehicles once had an important role in the daily operations on these farms.

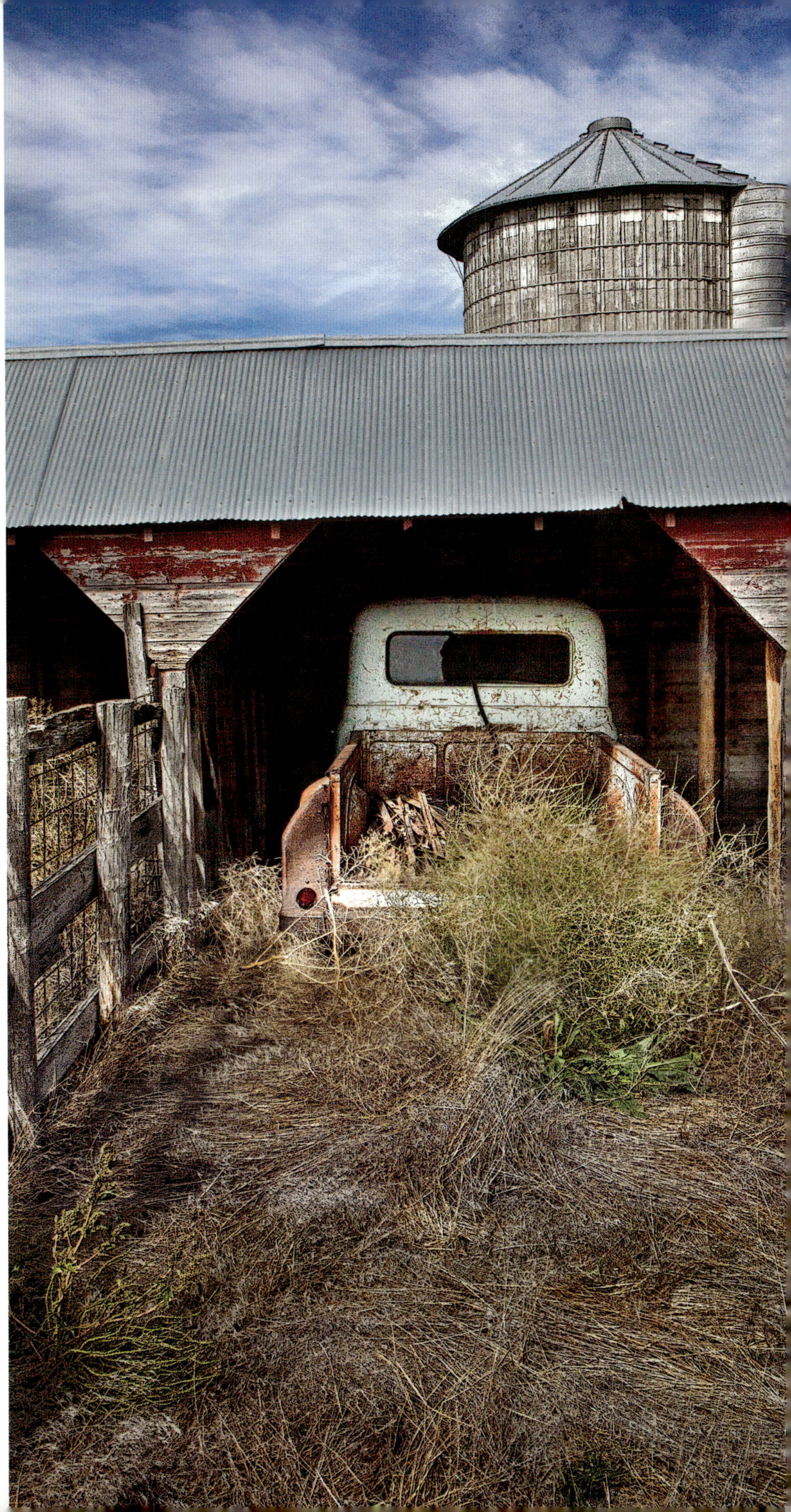

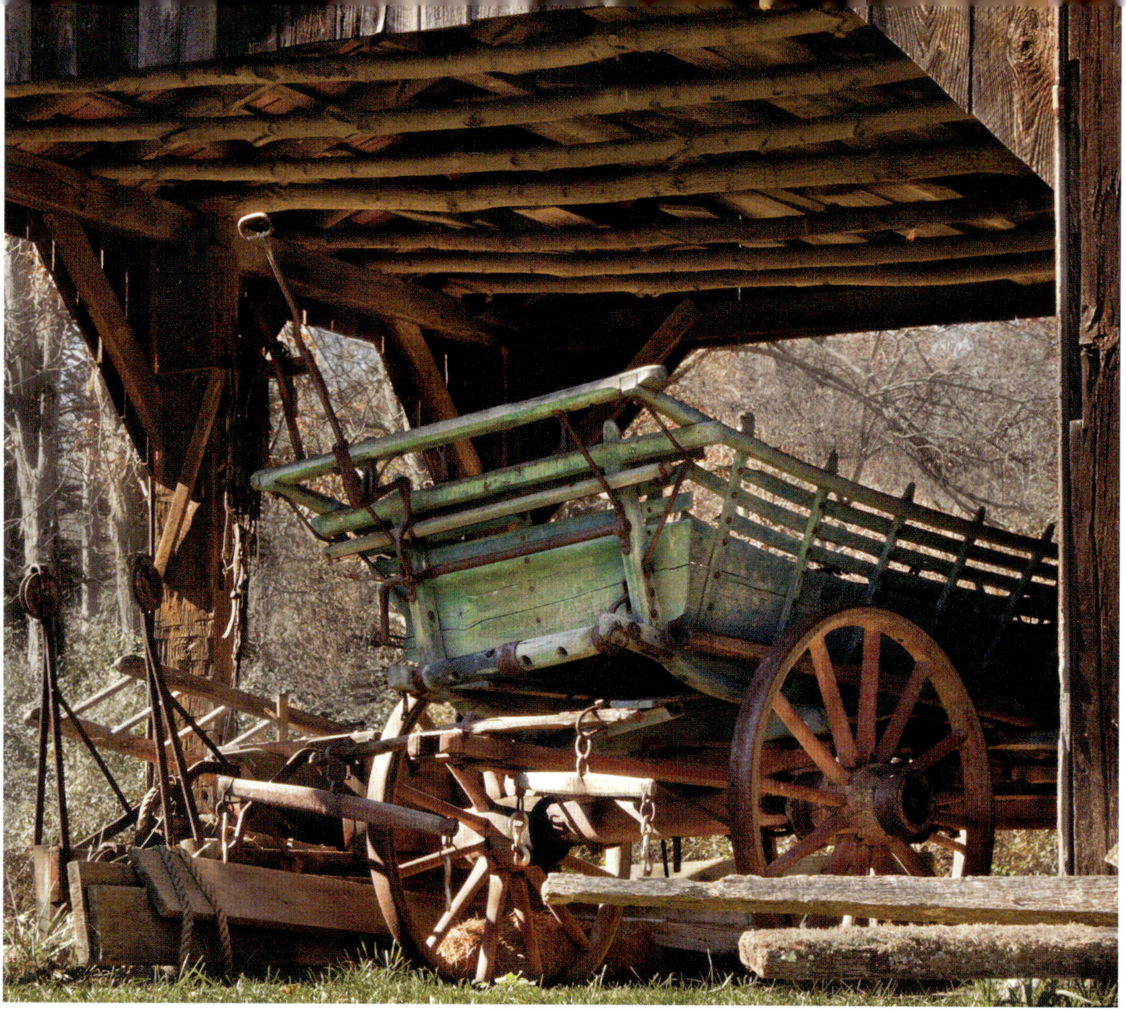

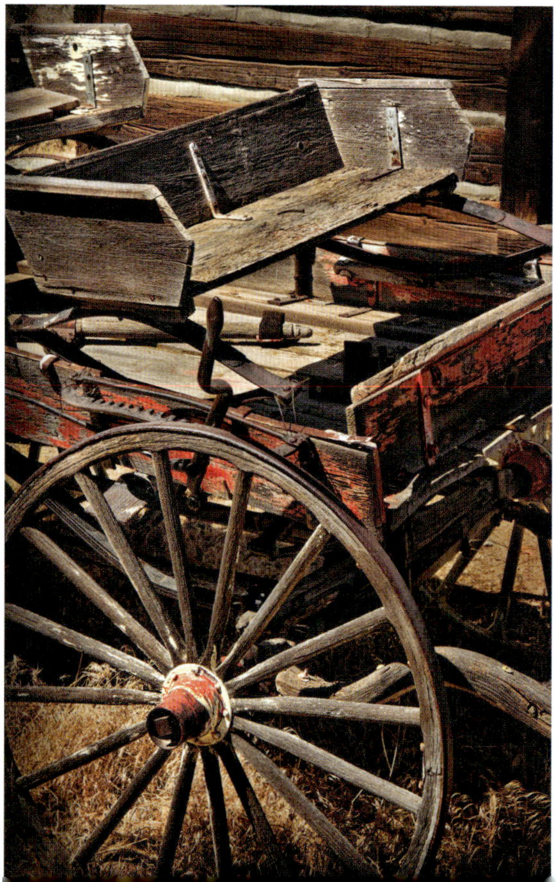

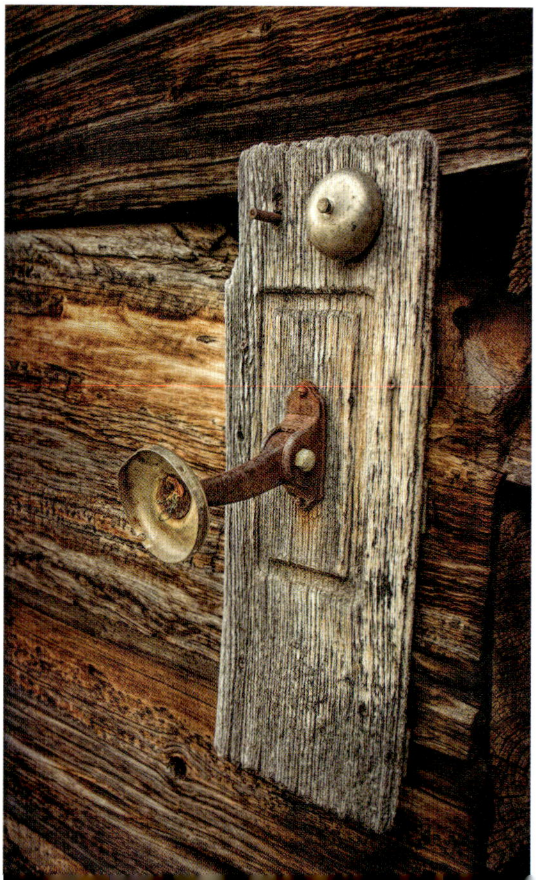

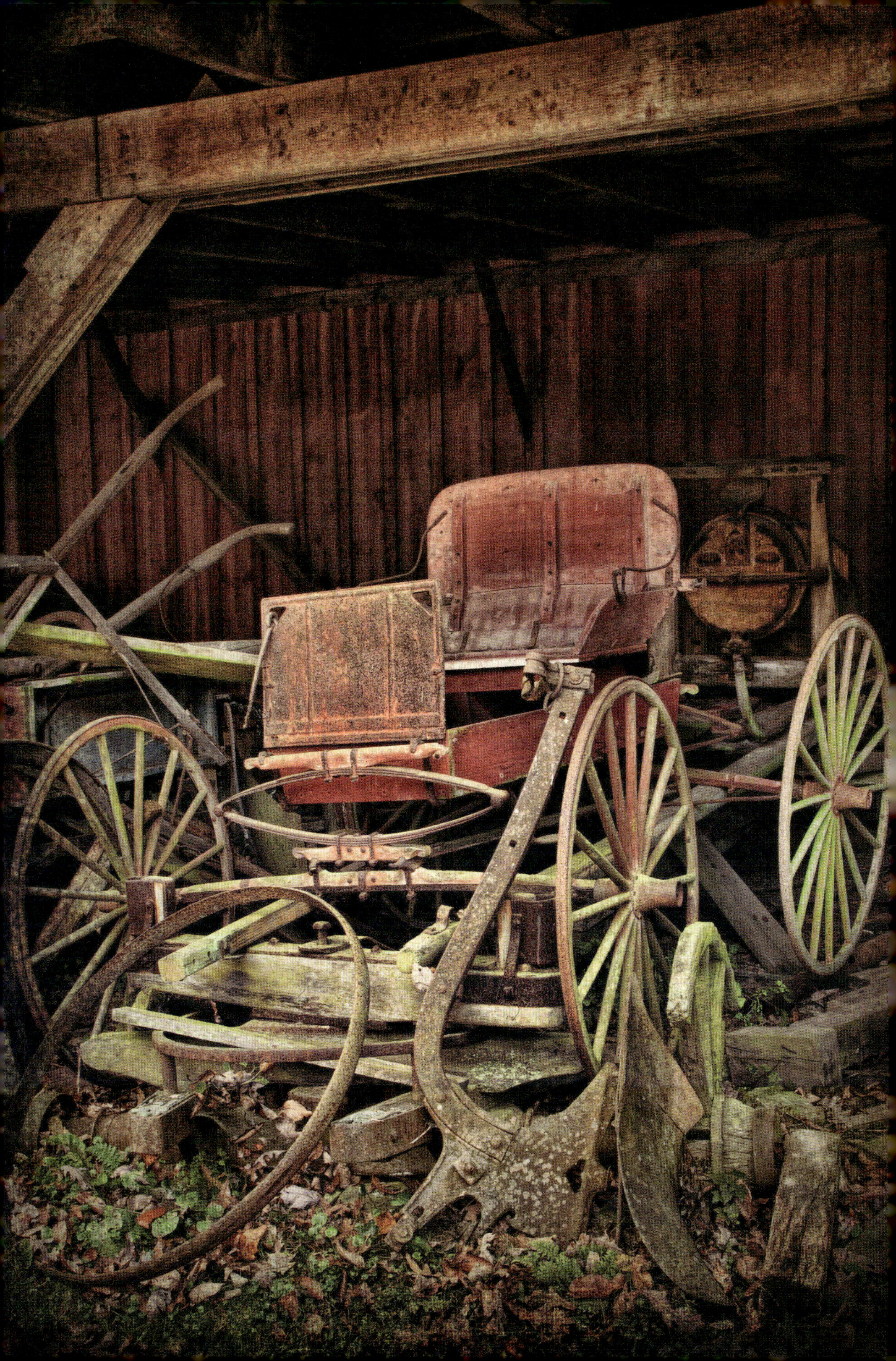

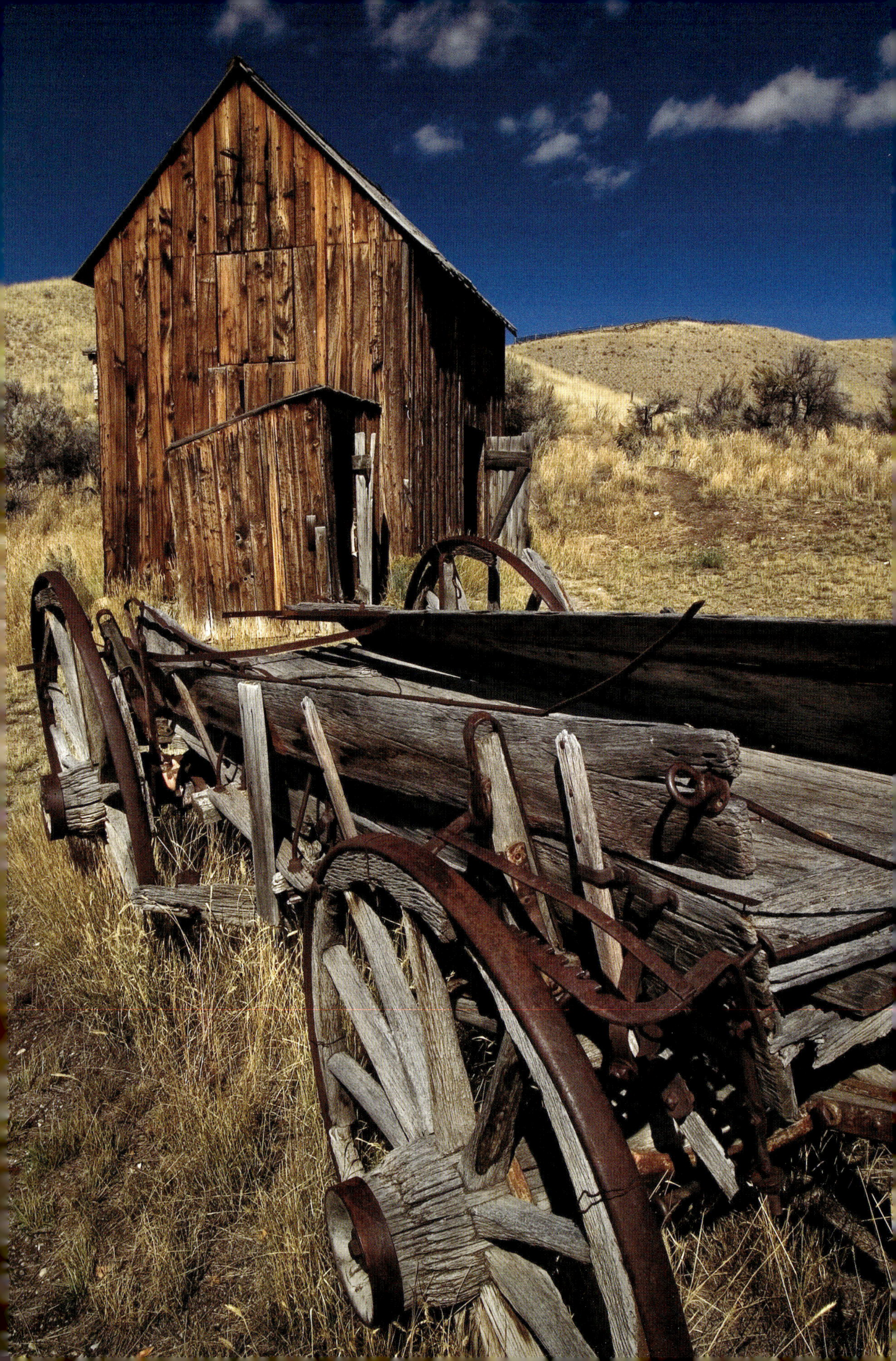

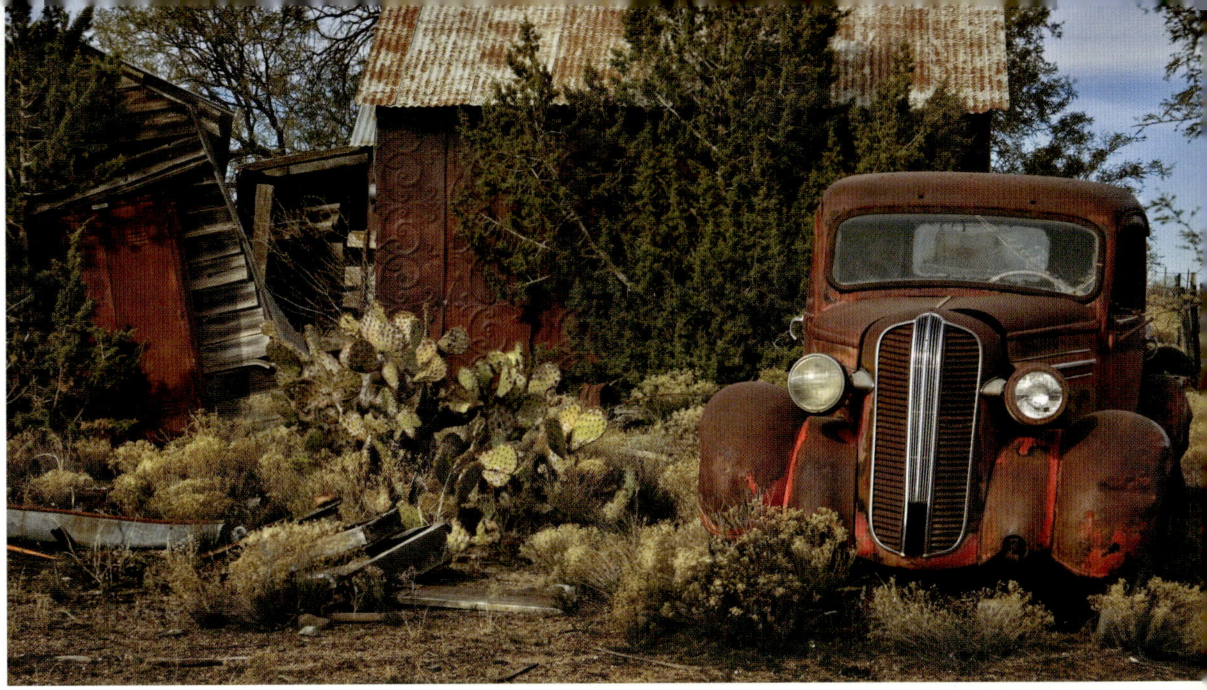

**previous page**—Prairie mountain barn. Montana.

**above**—*Rusted Through Time*. A vintage pickup truck in Cordes Station, Arizona.

**right**—A tool of another trade, tucked away in the back of a barn.

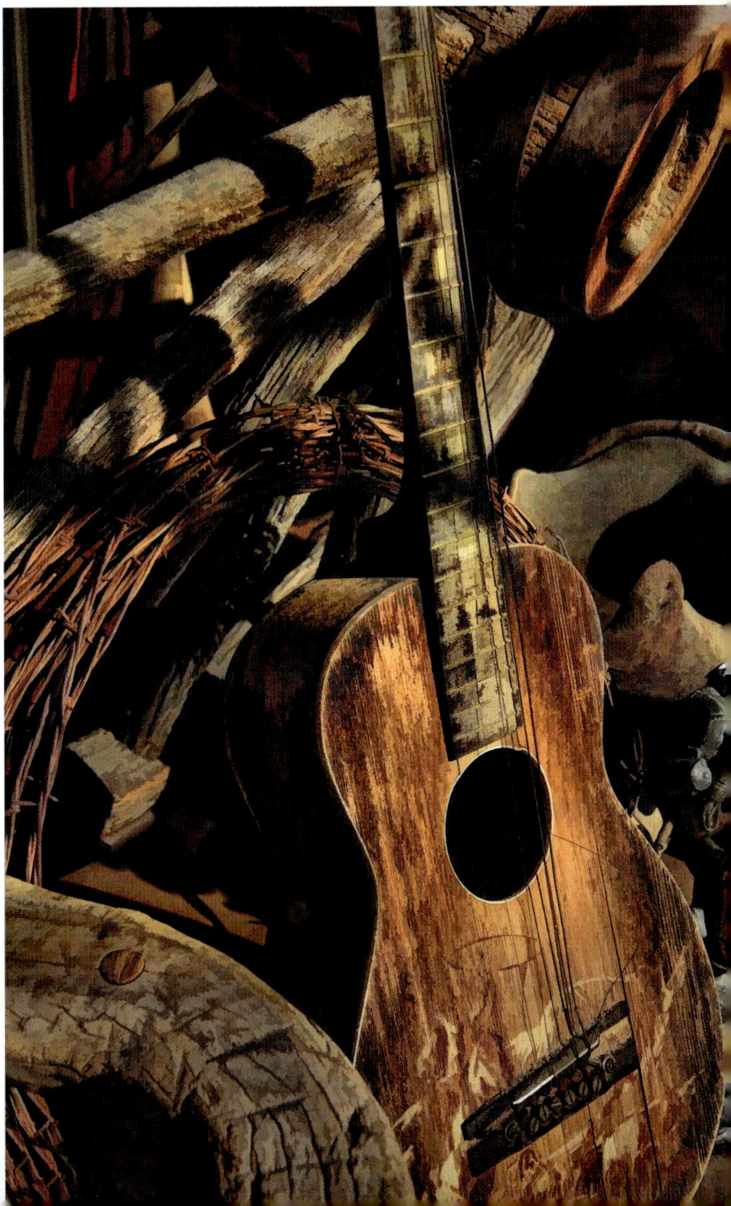

# Change of Seasons

The beauty of the rural landscape is a sight to see, no matter the time of year. In spring comes the first signs of new life; summer brings golden sun and harvests; fall provides a feast for the eyes—an abundance of color—and the crisp air of the season; and winter creates an almost audible silence that allows for rest and reflection. In this chapter, you'll find images of numerous barns and farms across America, photographed in spring, summer, winter, and fall.

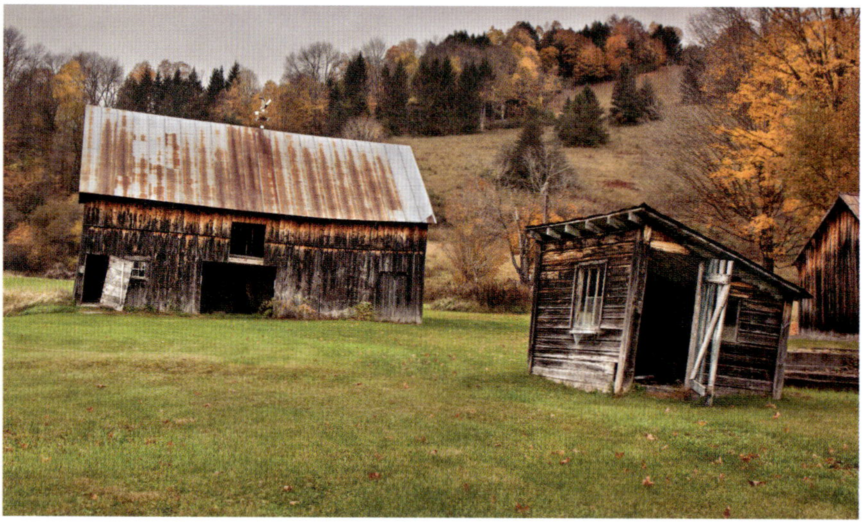

**left and following page**—Vermont..

**below**—New Hampshire.

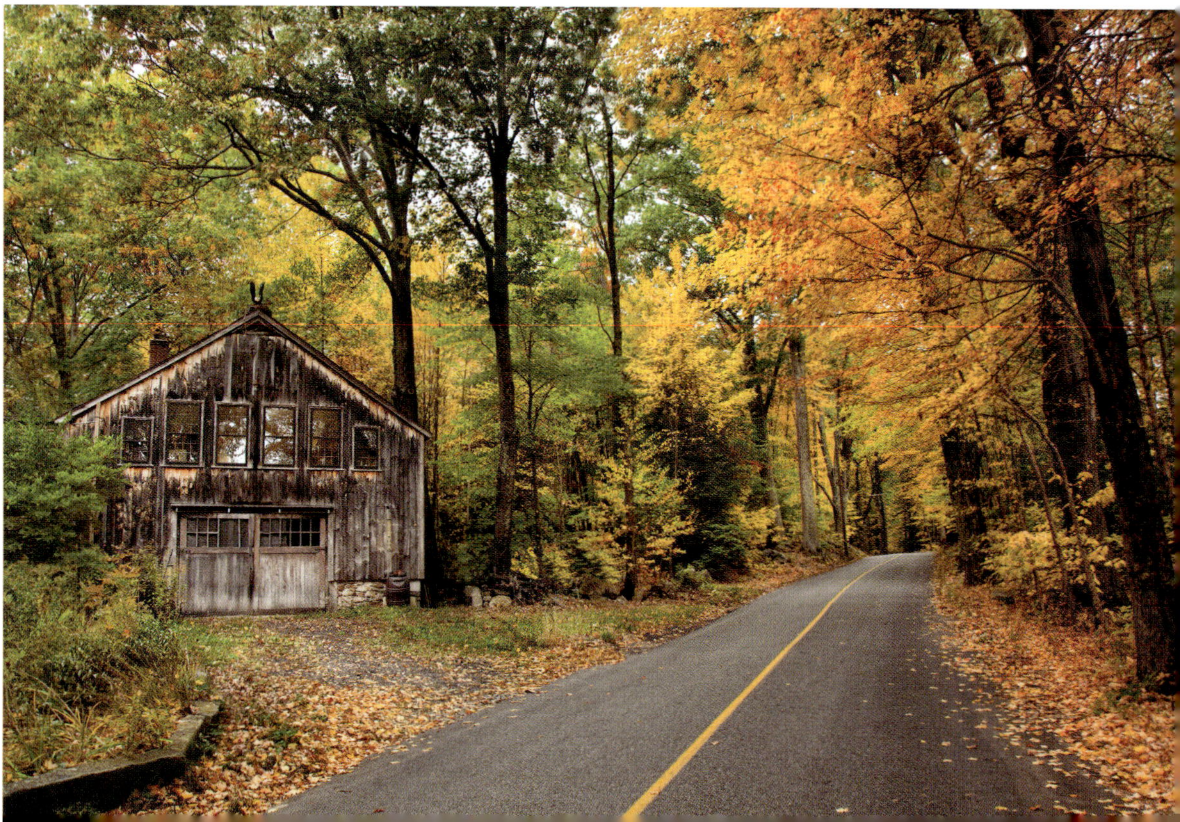

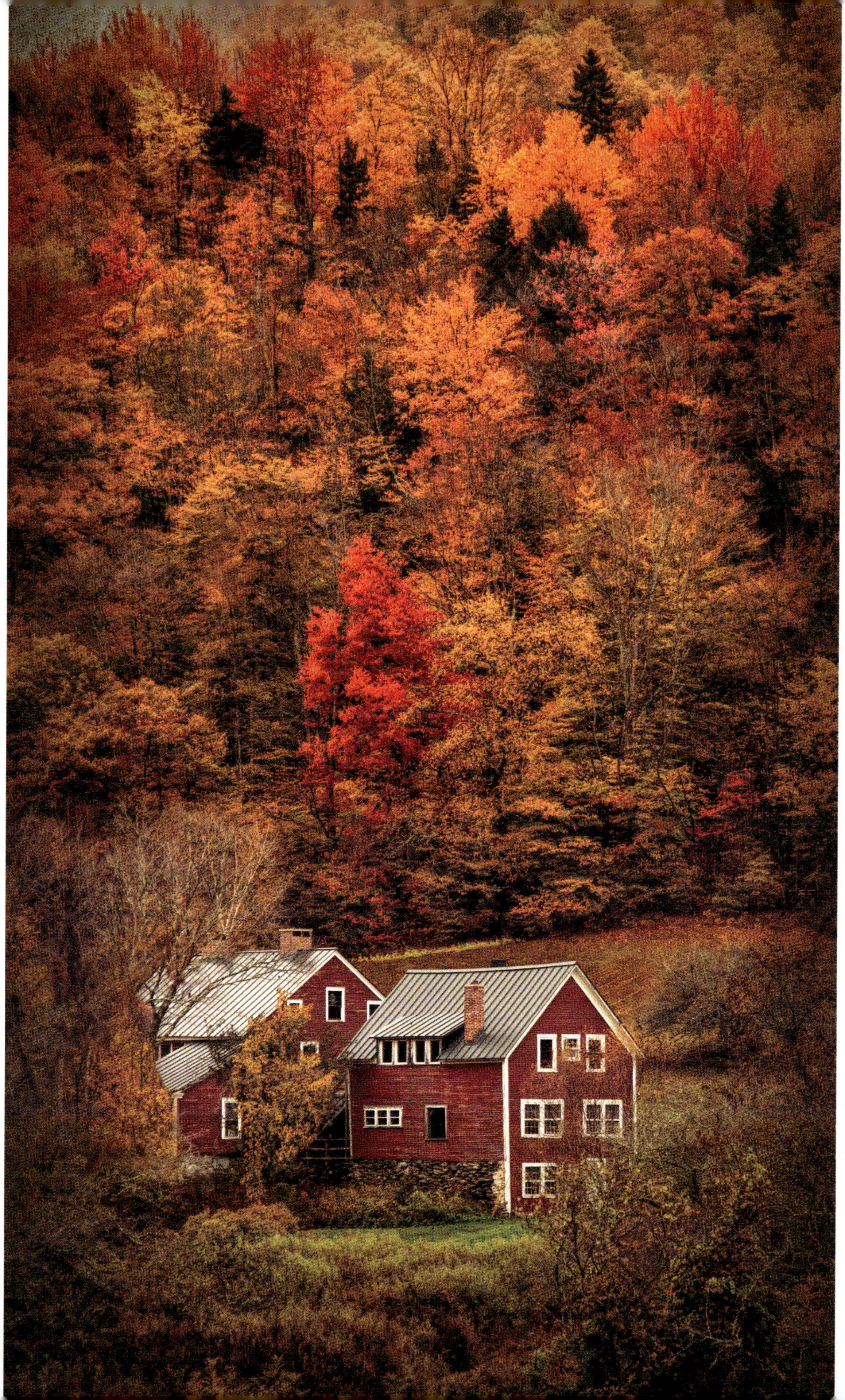

"My favorite color is October."

—Jeffrey Steccato

**top and bottom**—Amish farms in Pennsylvania.

**following page, top**—Virginia.

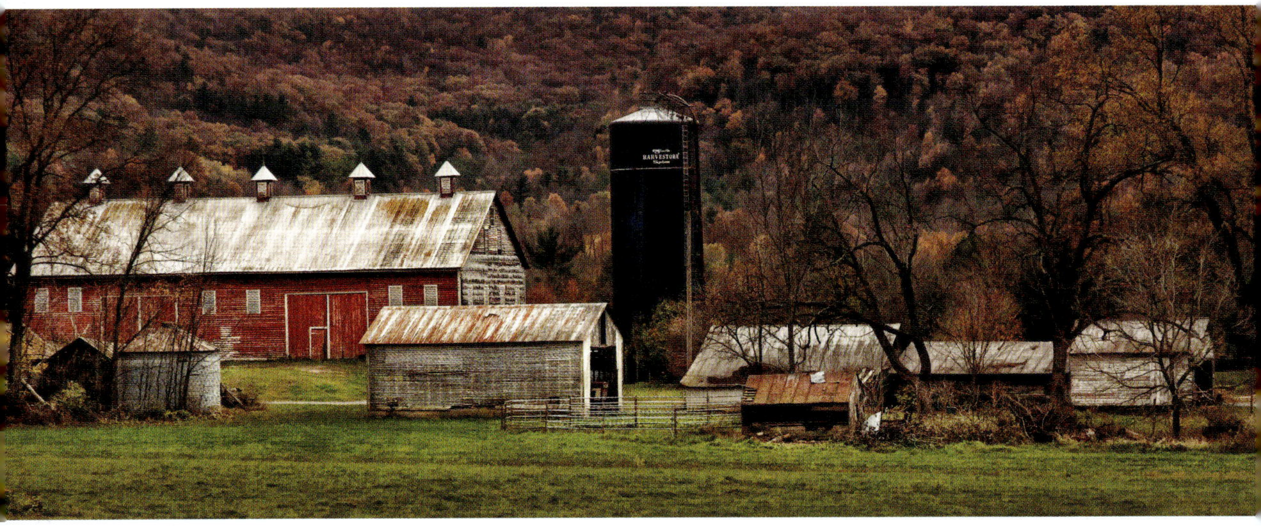

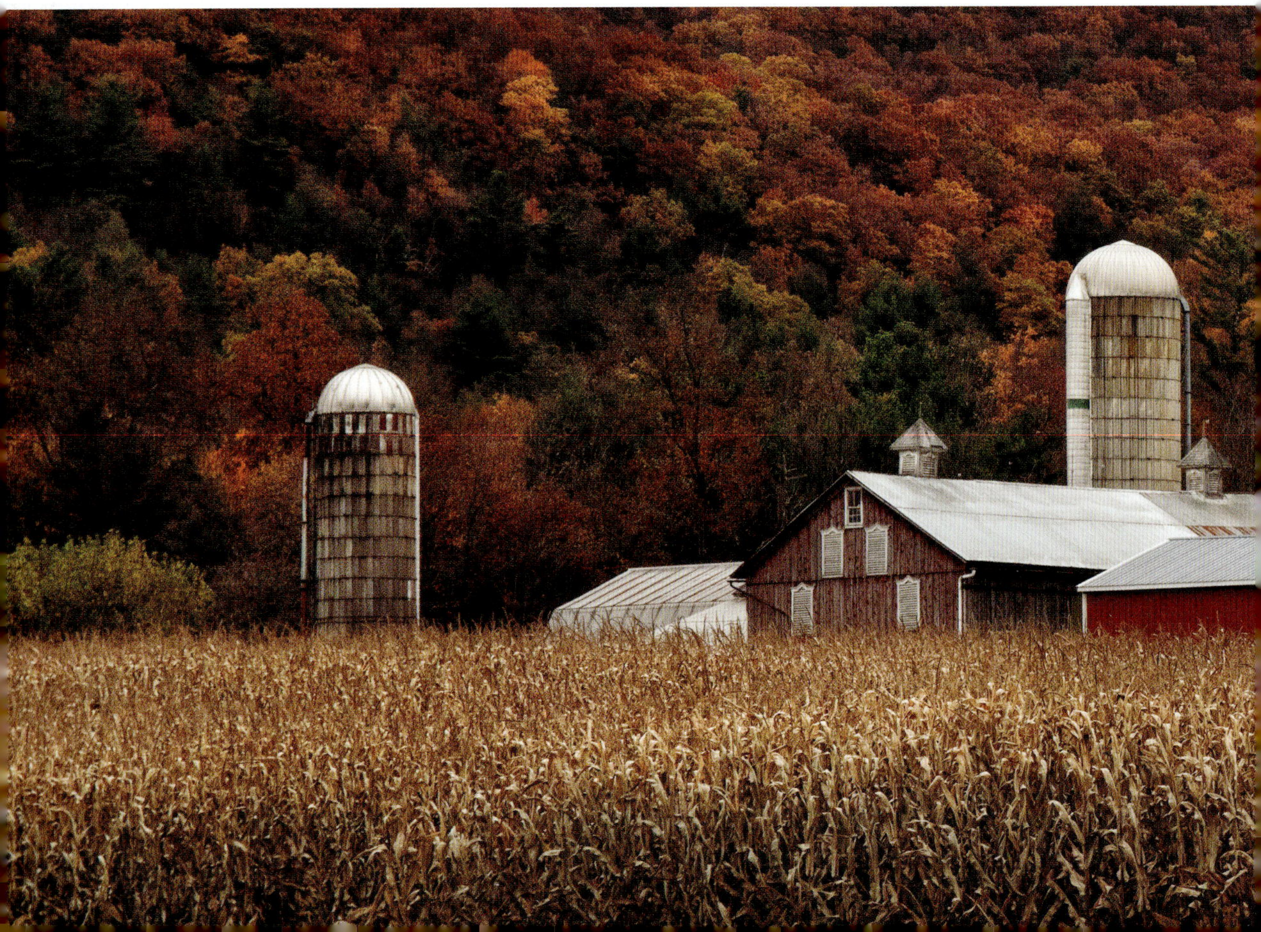

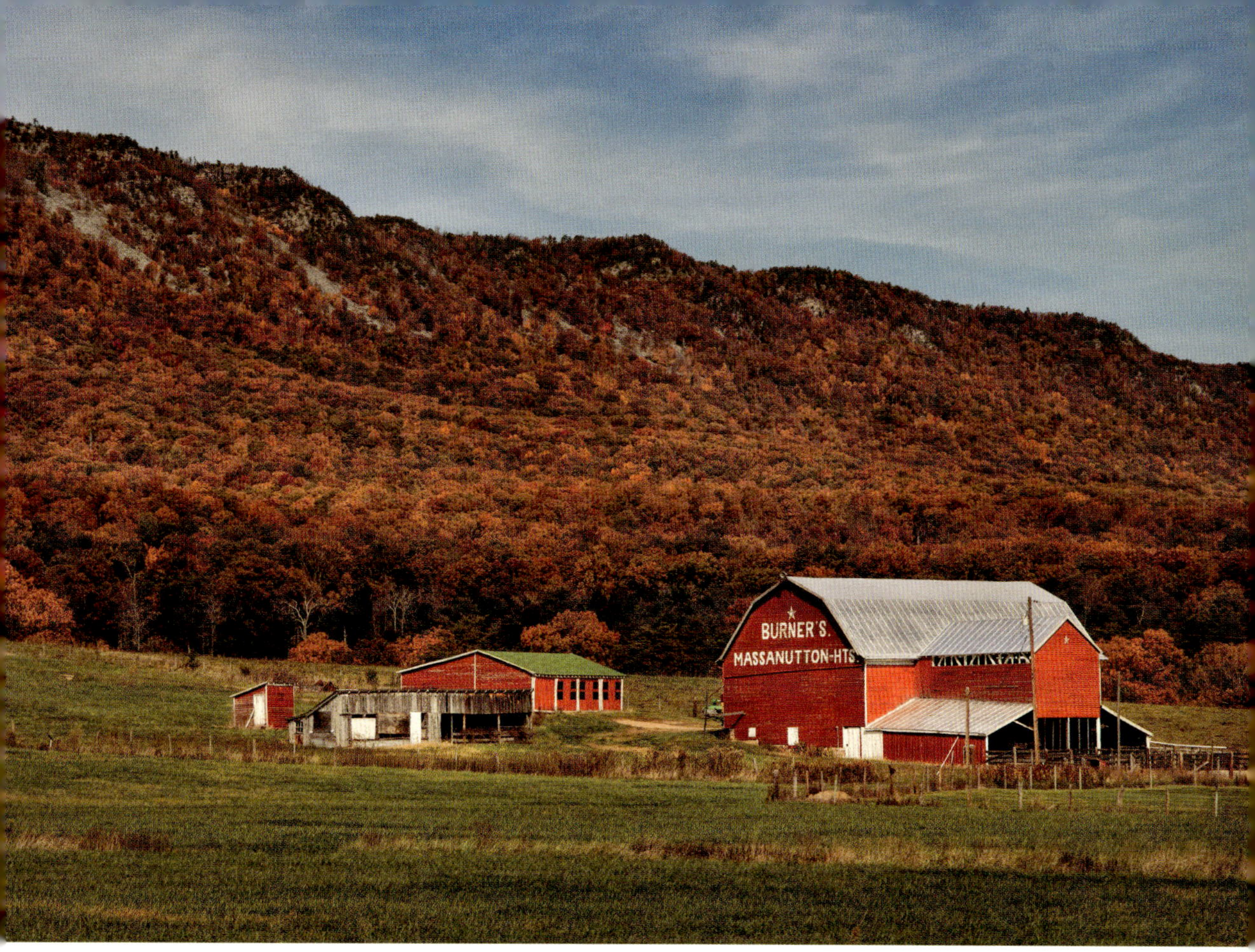
BURNER'S
MASSANUTTON-HTS

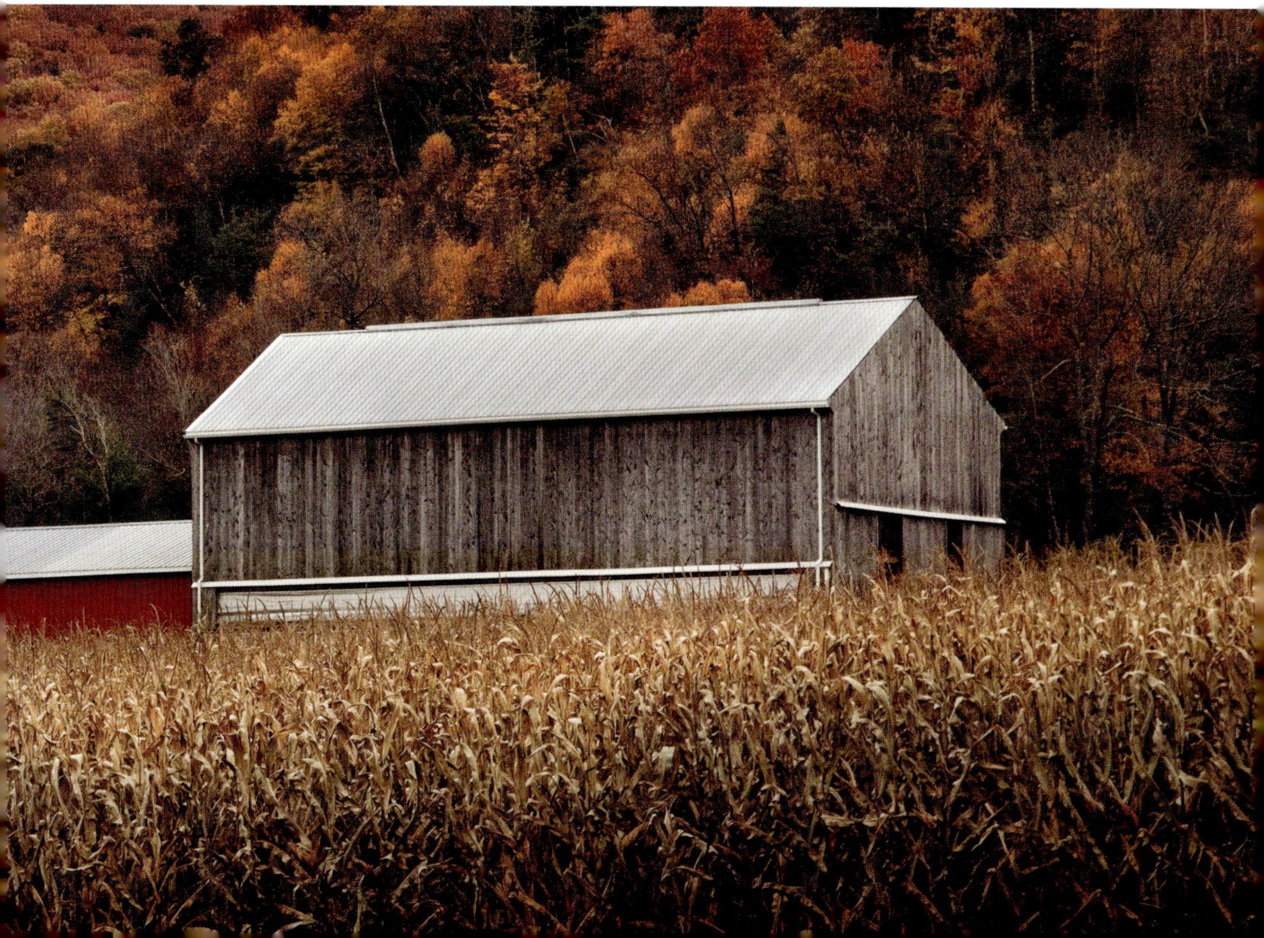

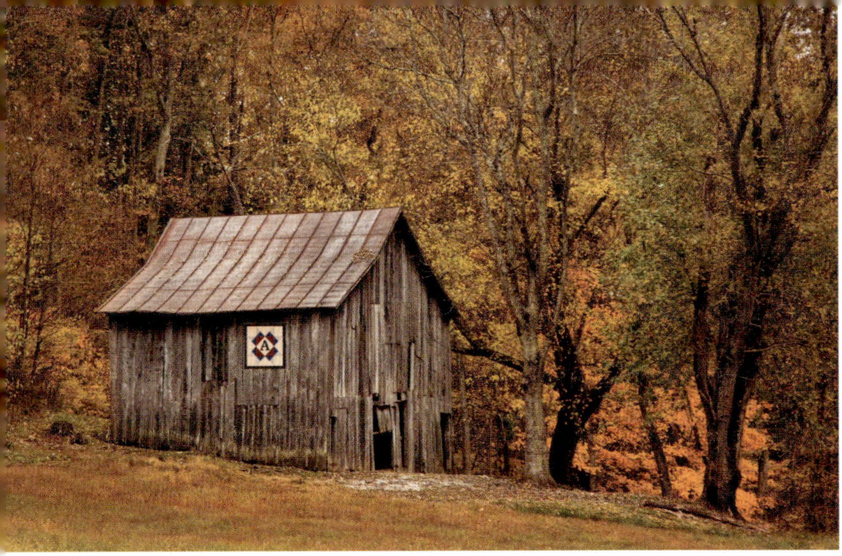

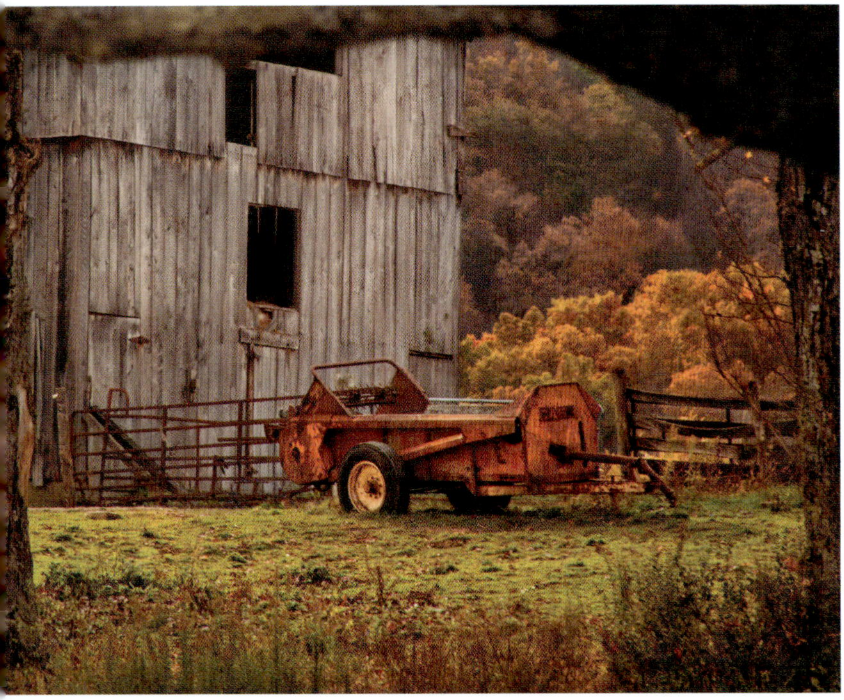

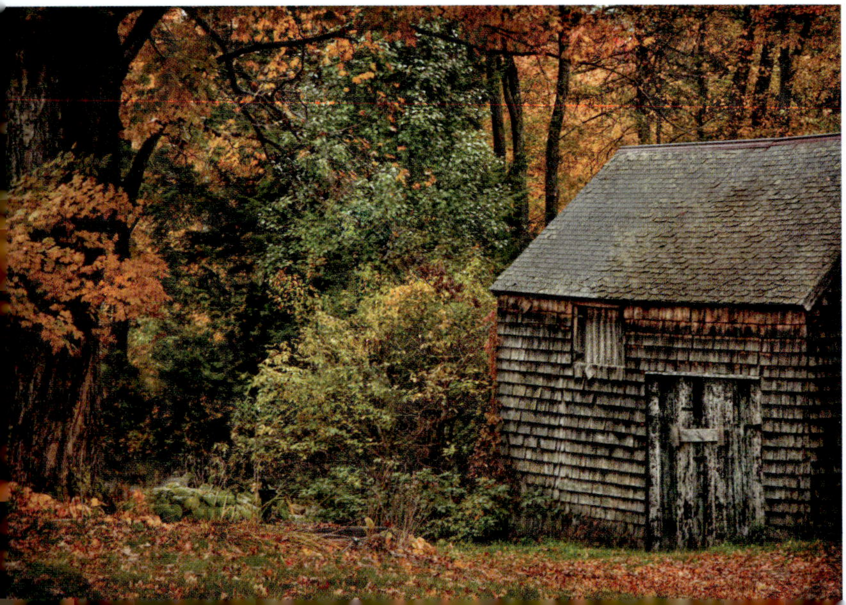

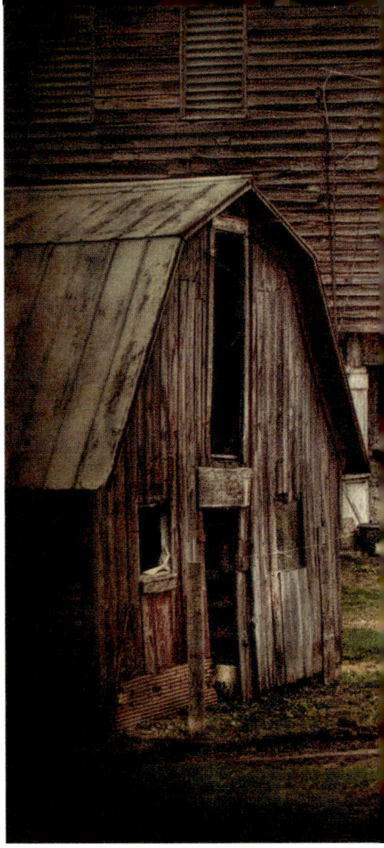

**top left**—Virginia.

**center left**—West Virginia.

**bottom left**—New Hampshire.

**above**—Pennsylvania.

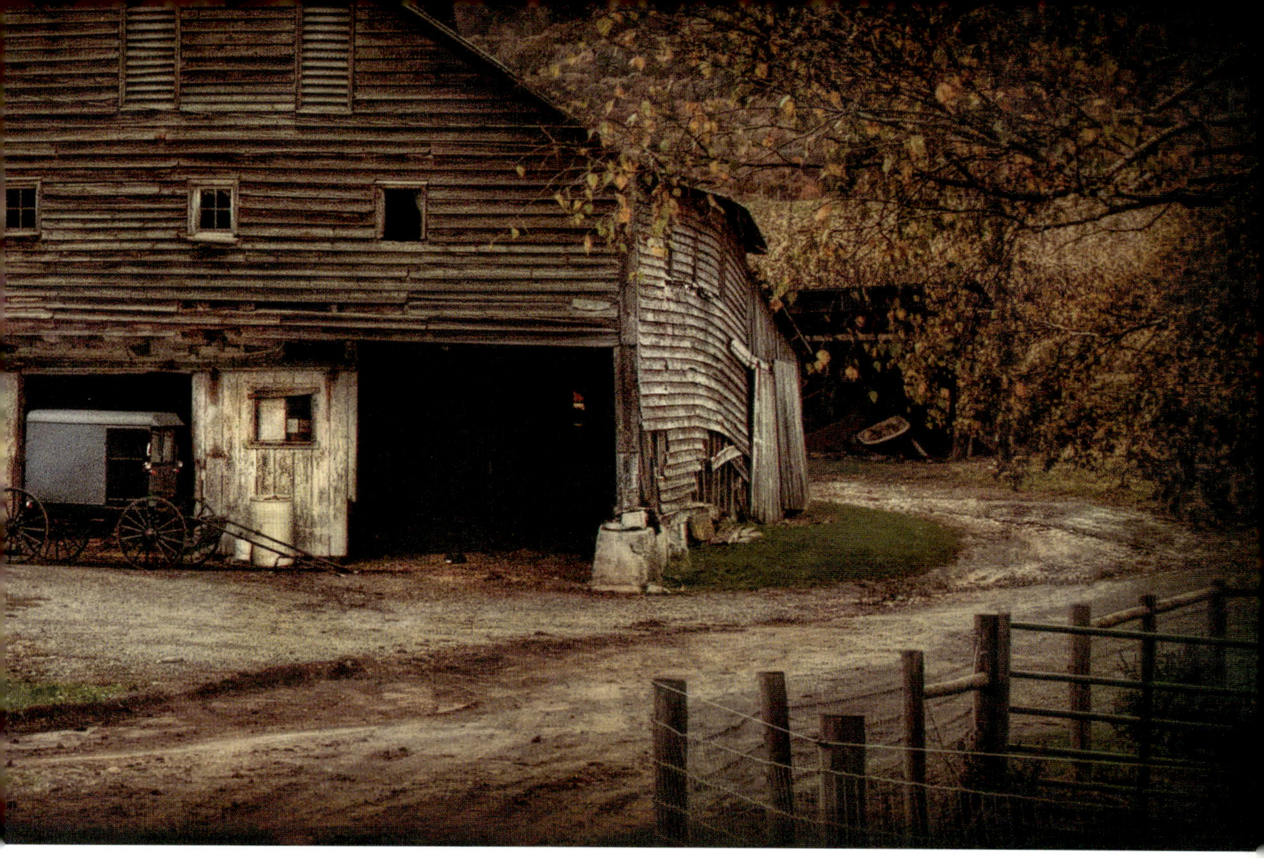

"If you can't feel what you're looking at, then you're never going to get others to feel anything when they look at your pictures."

—Don McCullen

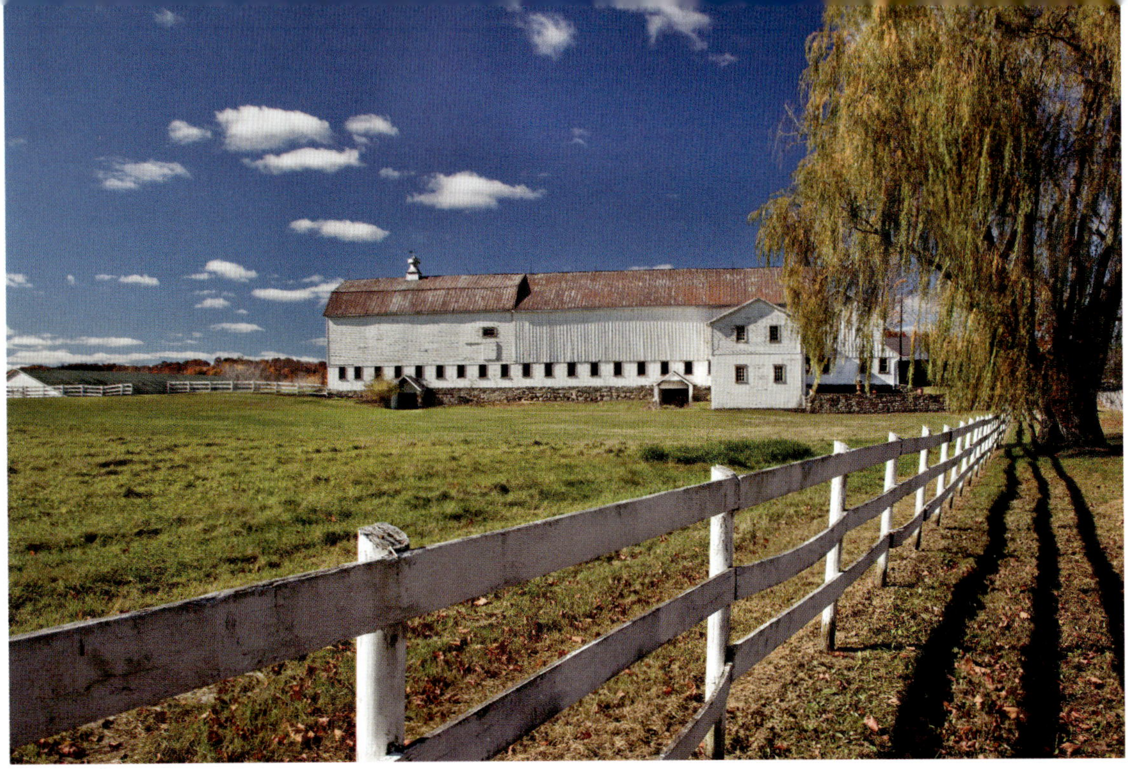

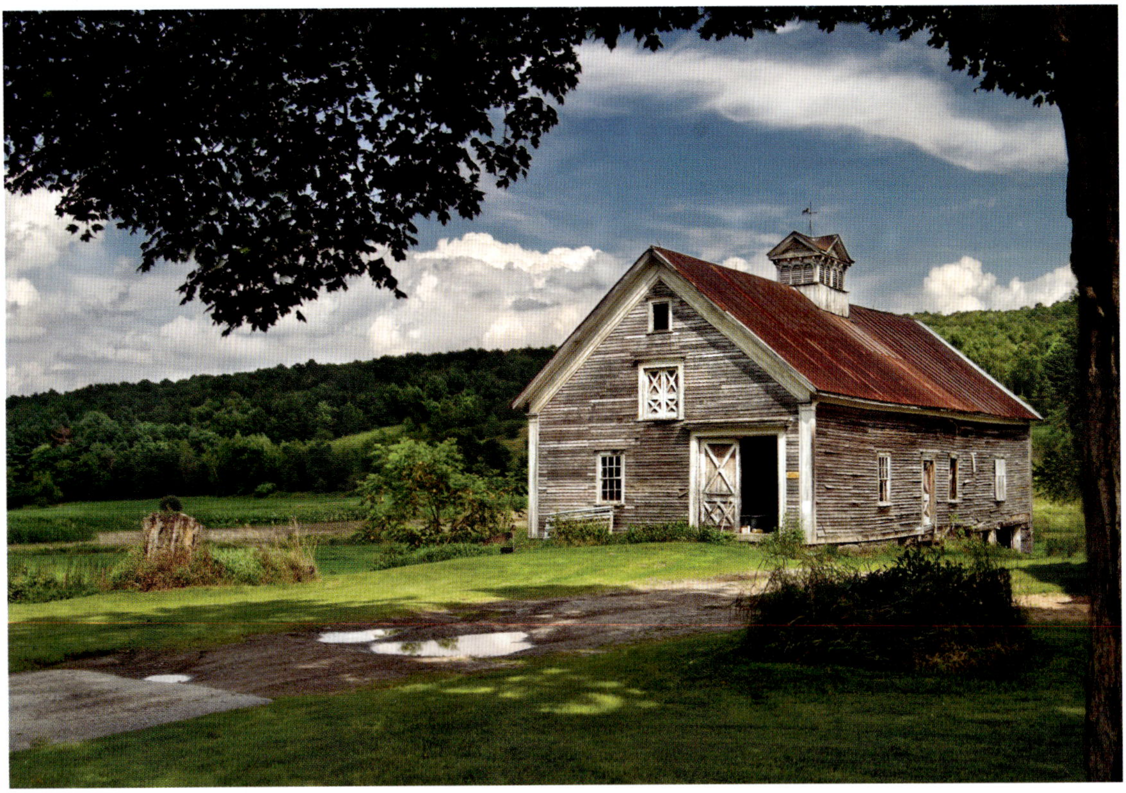

**top**—*Oceane Vineyards.* Chester, New York.

**bottom**—Connecticut.

**following page**—Massachusetts.

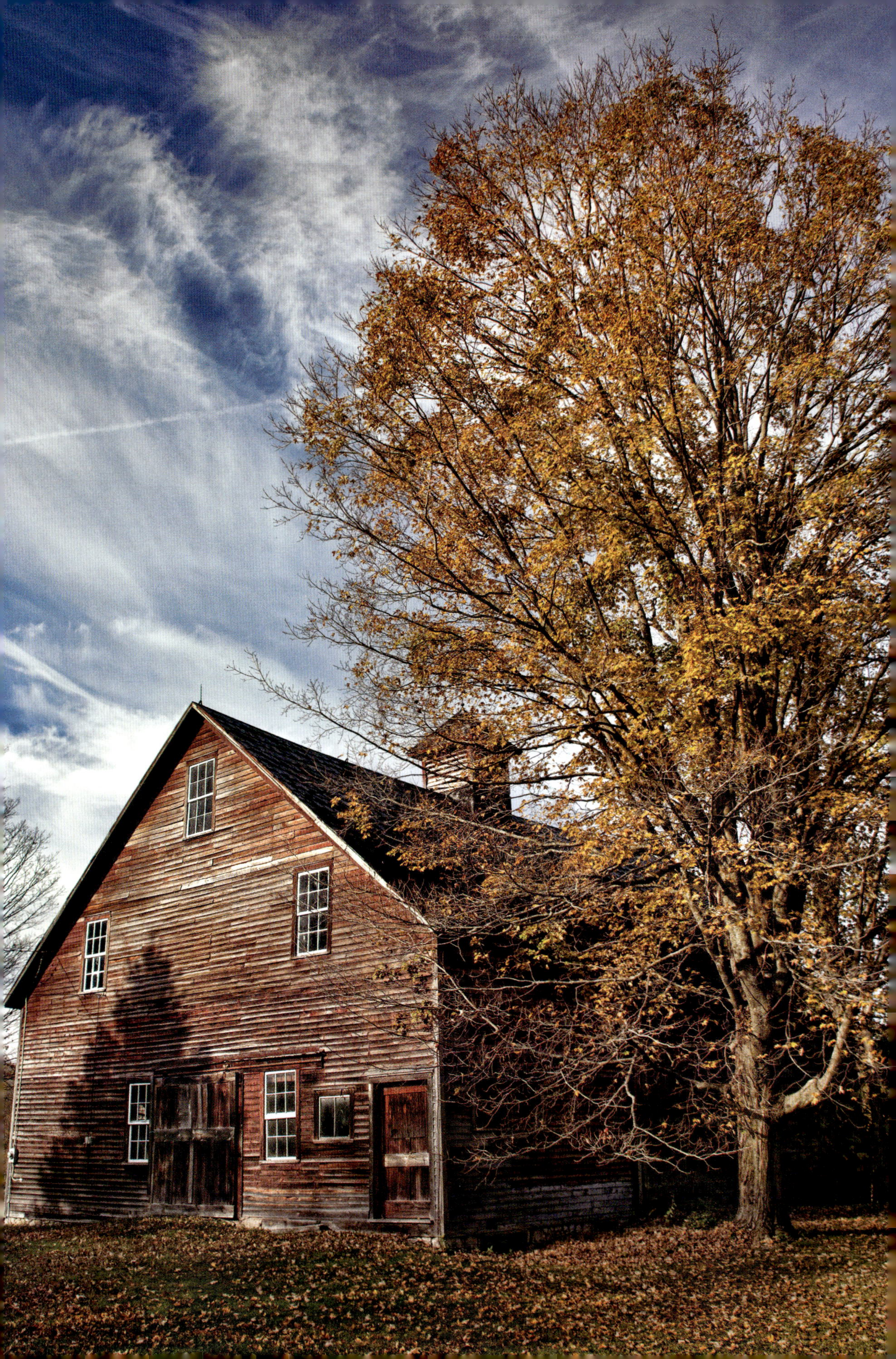

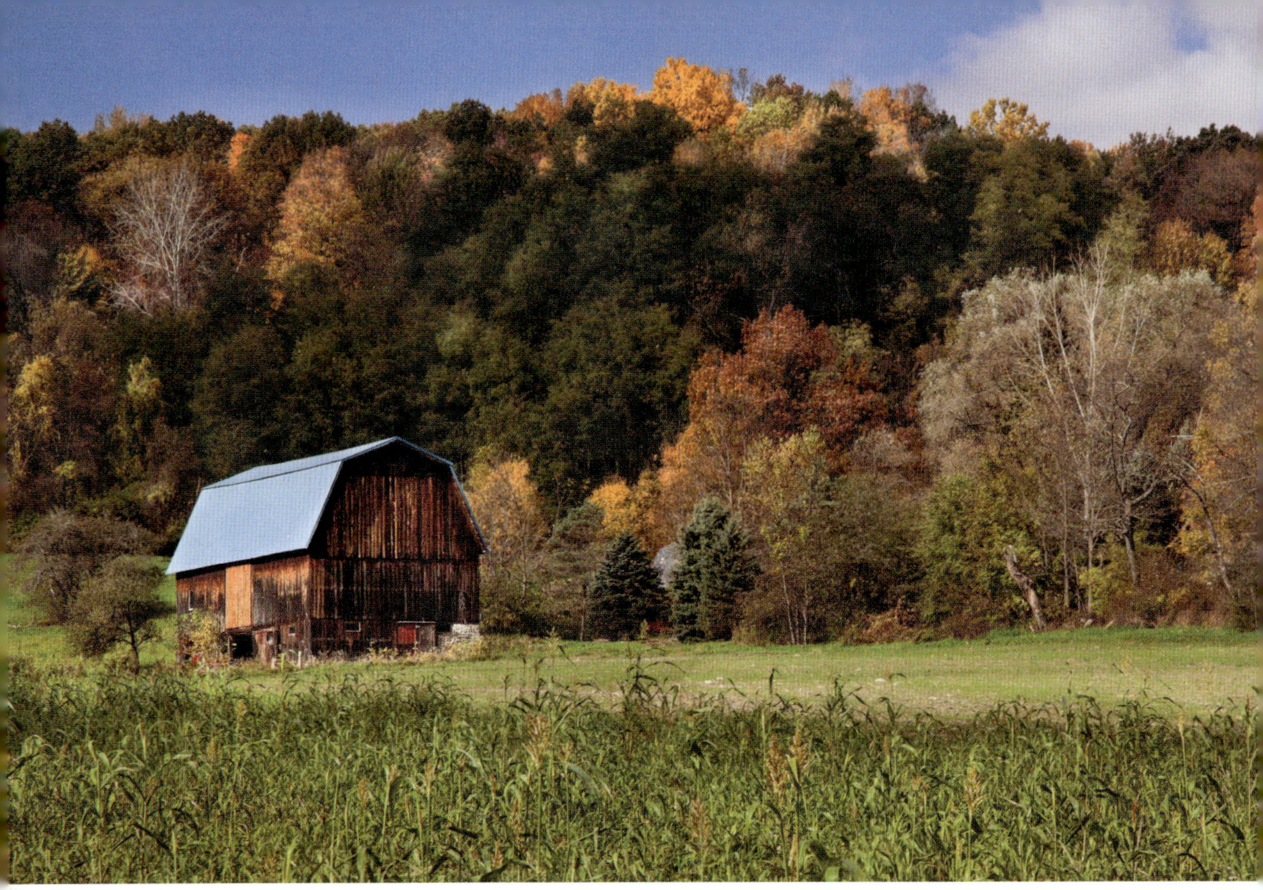

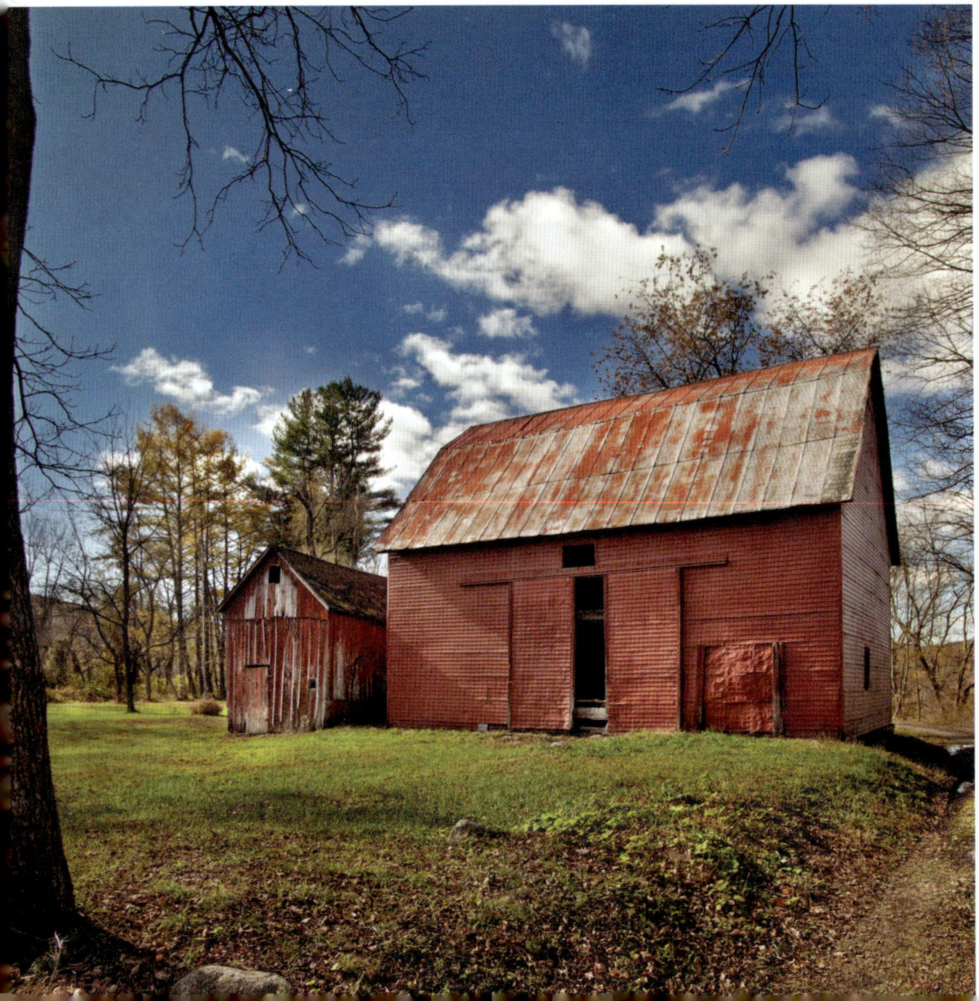

**above**—New York.

**left**—New Jersey.

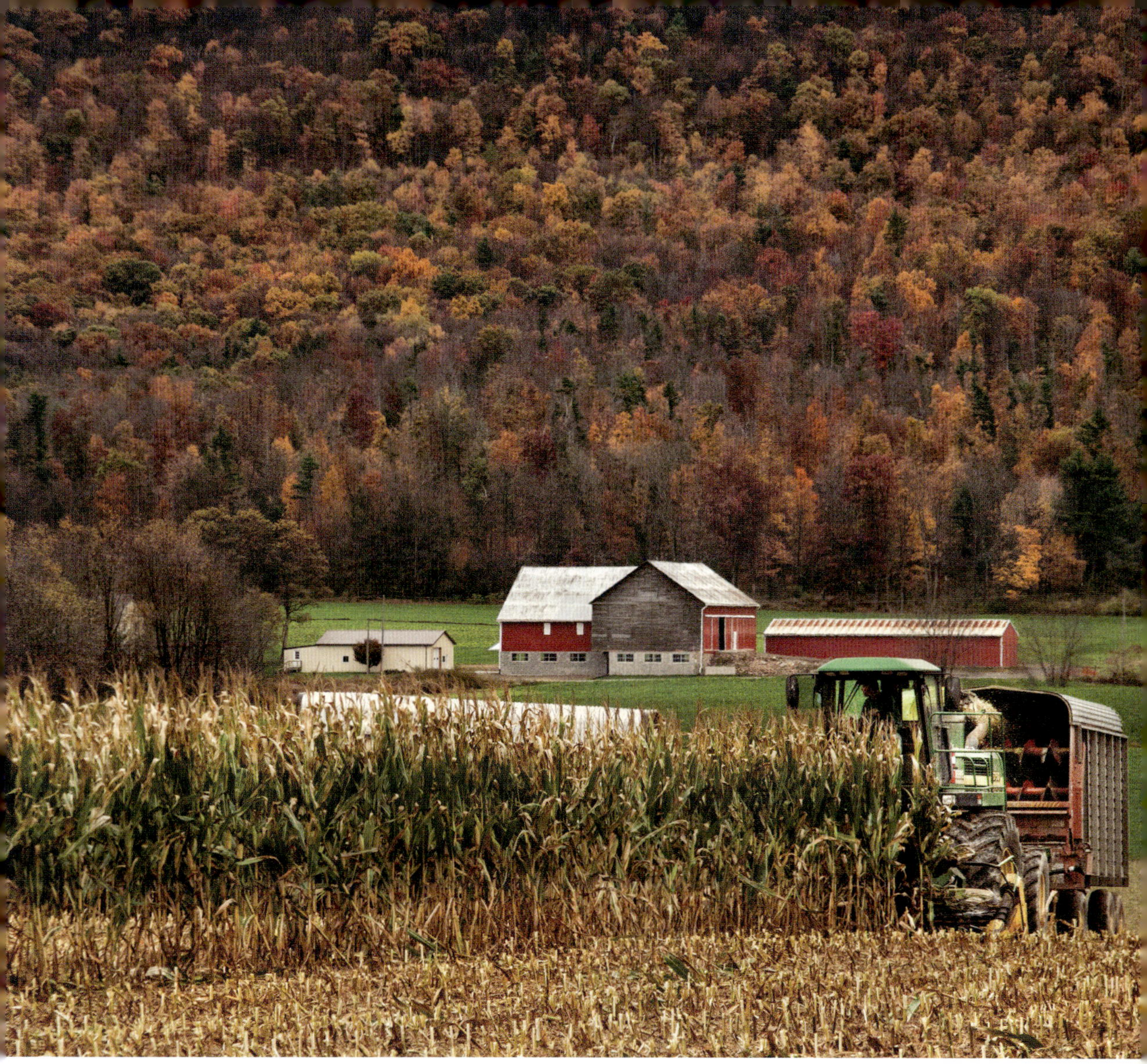

above—
Pennsylvania.

bottom left—
New York.

bottom right—
Virginia.

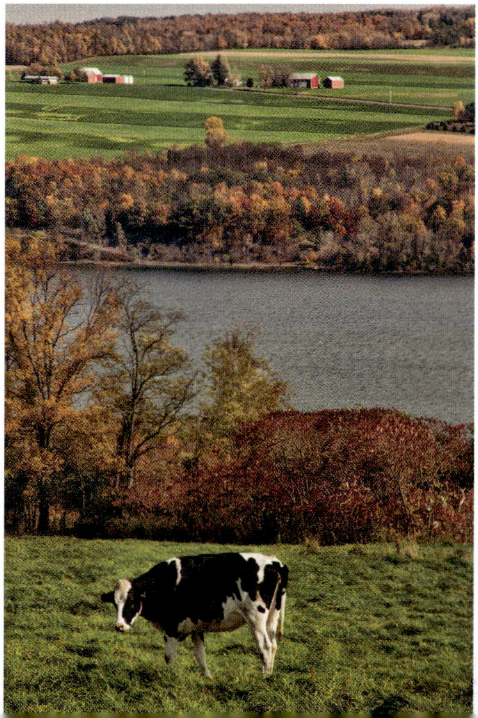

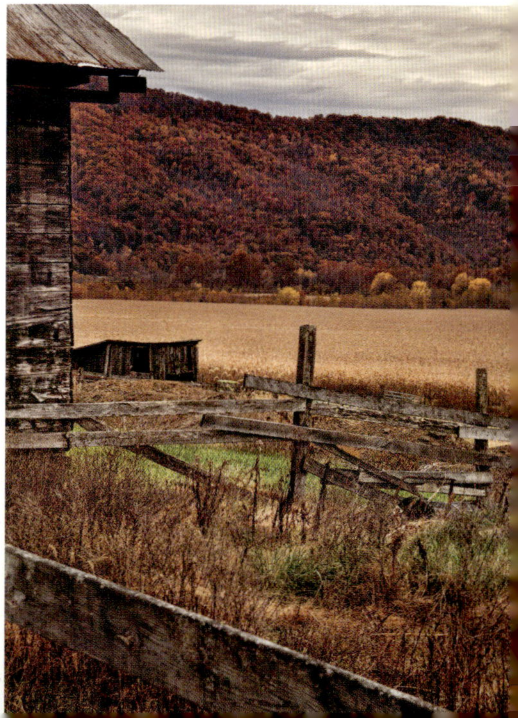

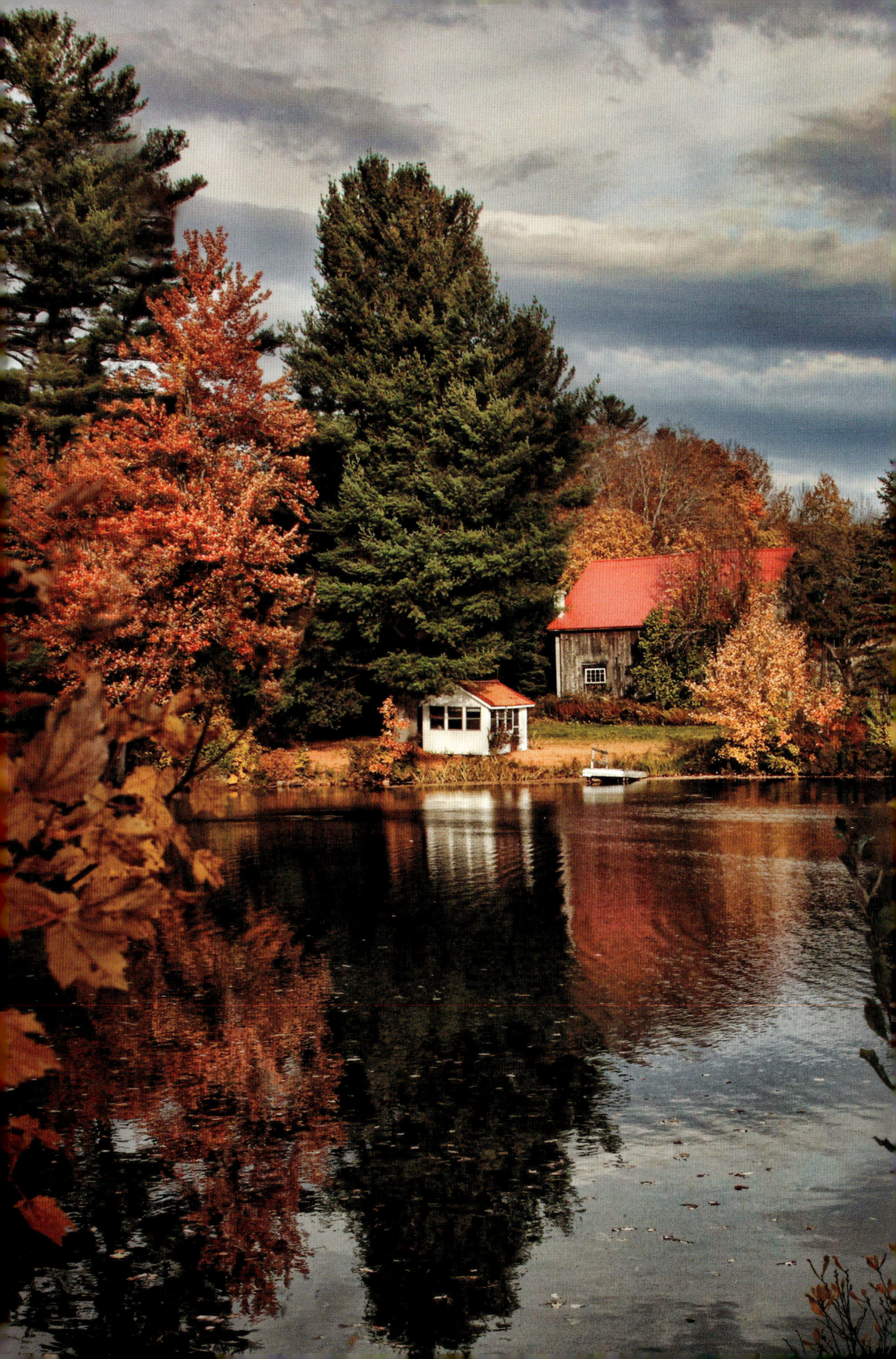

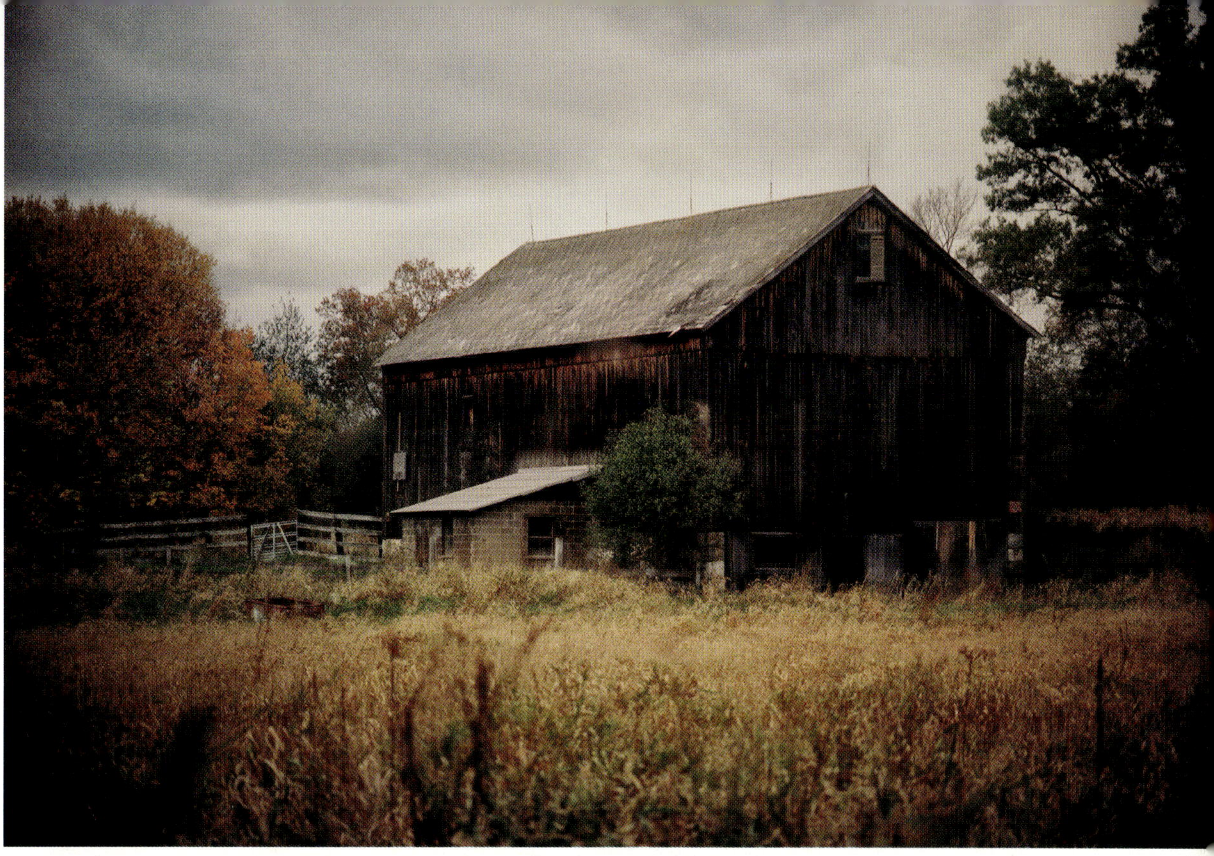

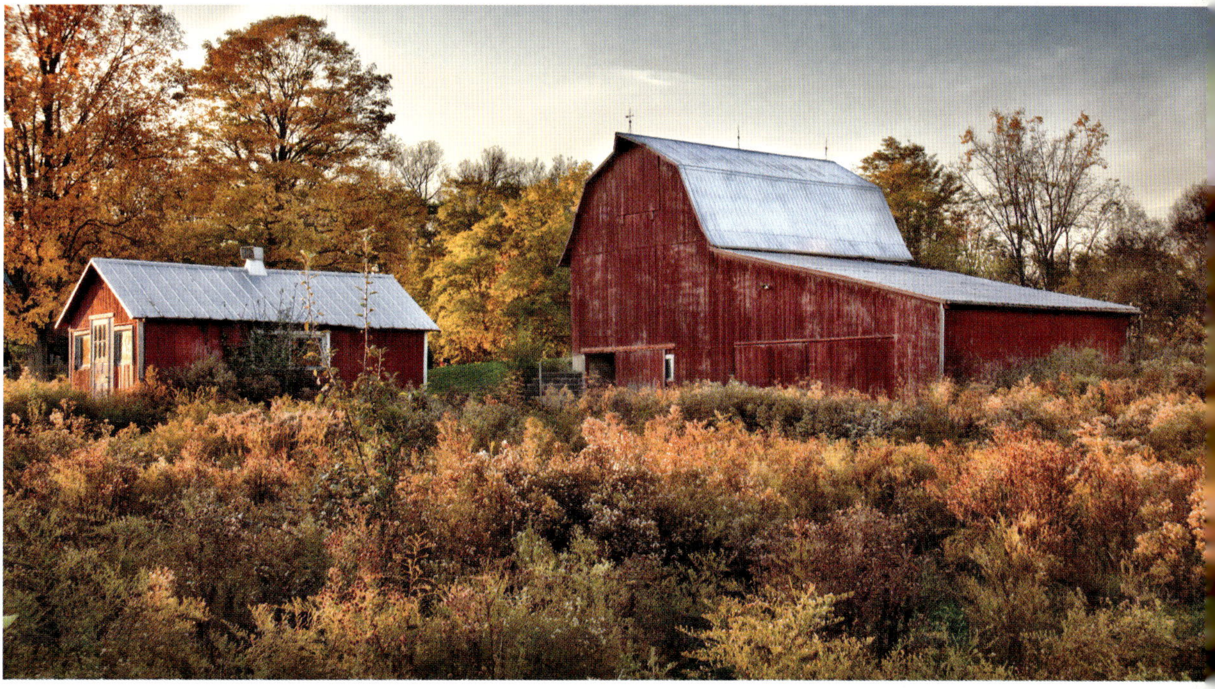

**previous page—**Reflections on the past. New Hampshire.

**top—**New York. Photographed on film.

**bottom—**New York.

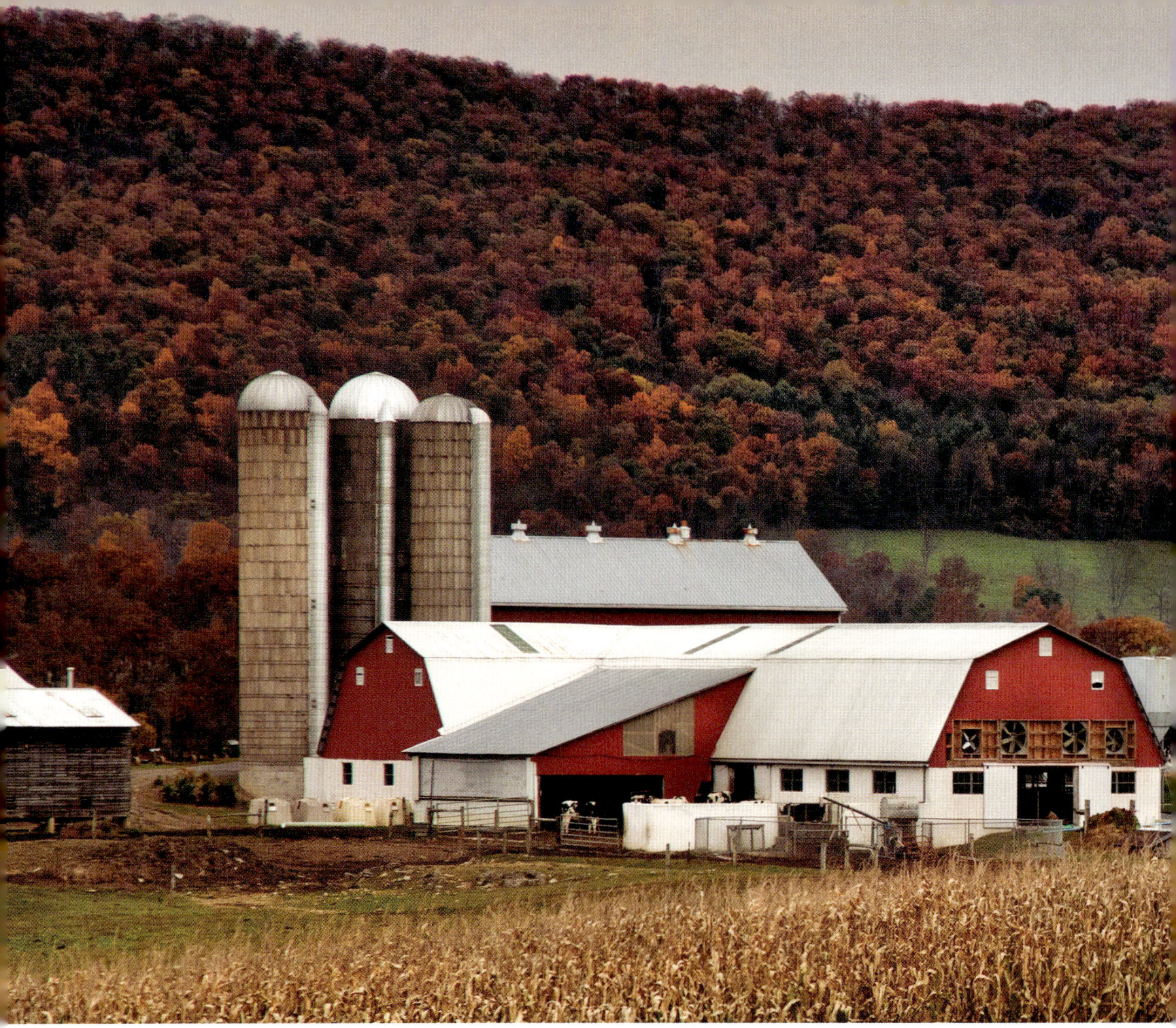

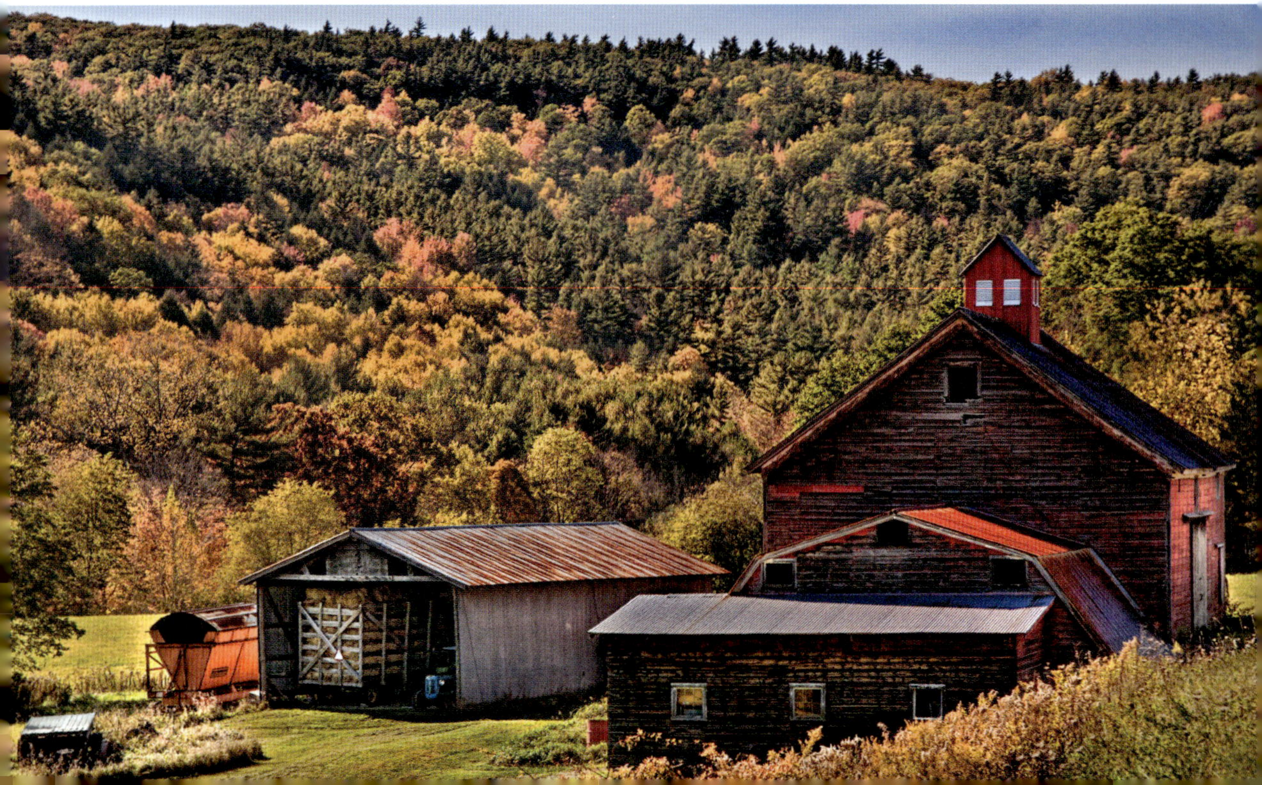

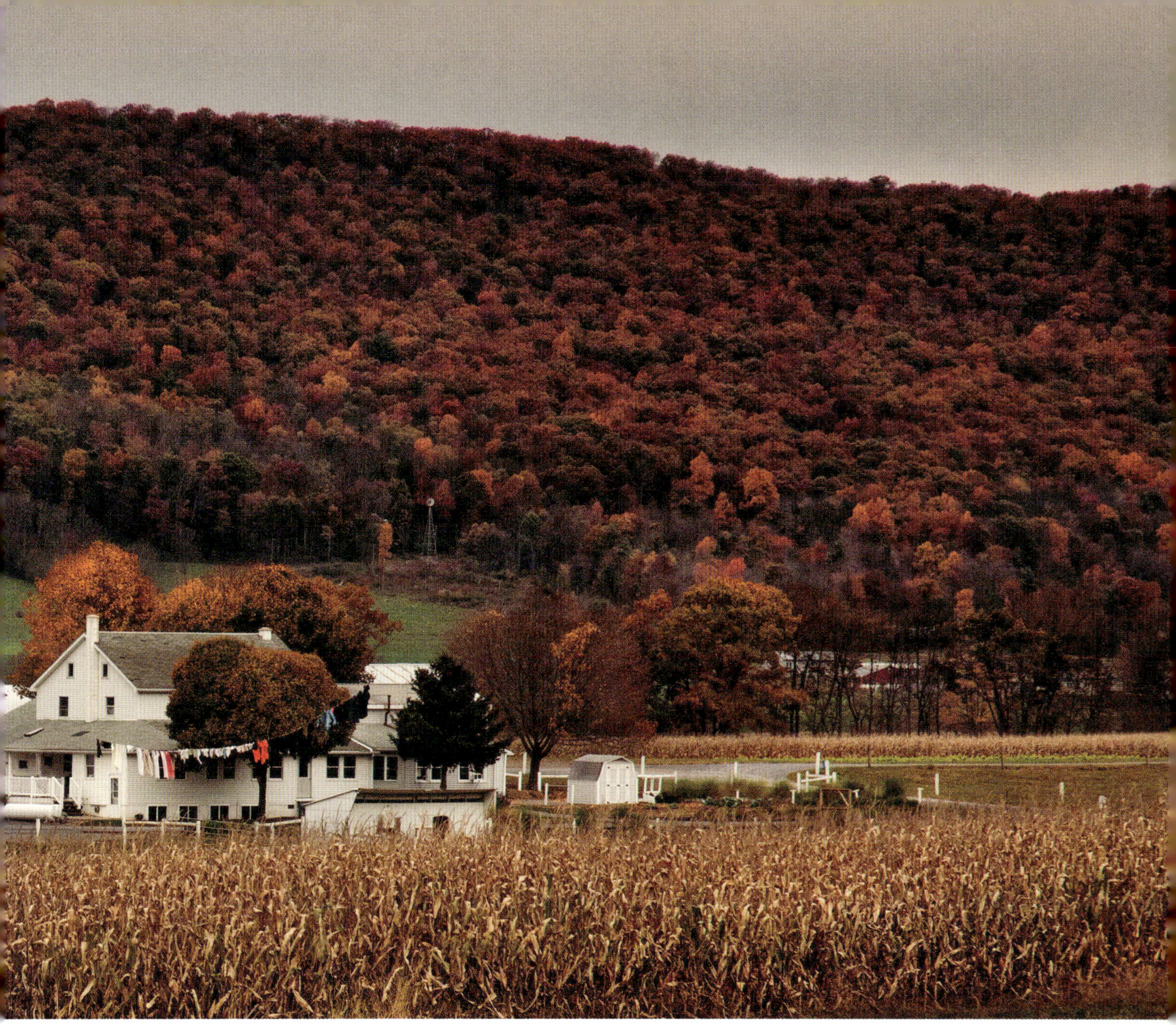

**above**—Amish farm. Pennsylvania.

**previous page, bottom**—*Yesterday's Harvest*. Massachusetts.

**"There are always two people in every picture: the photographer and the viewer."**

**—Ansel Adams**

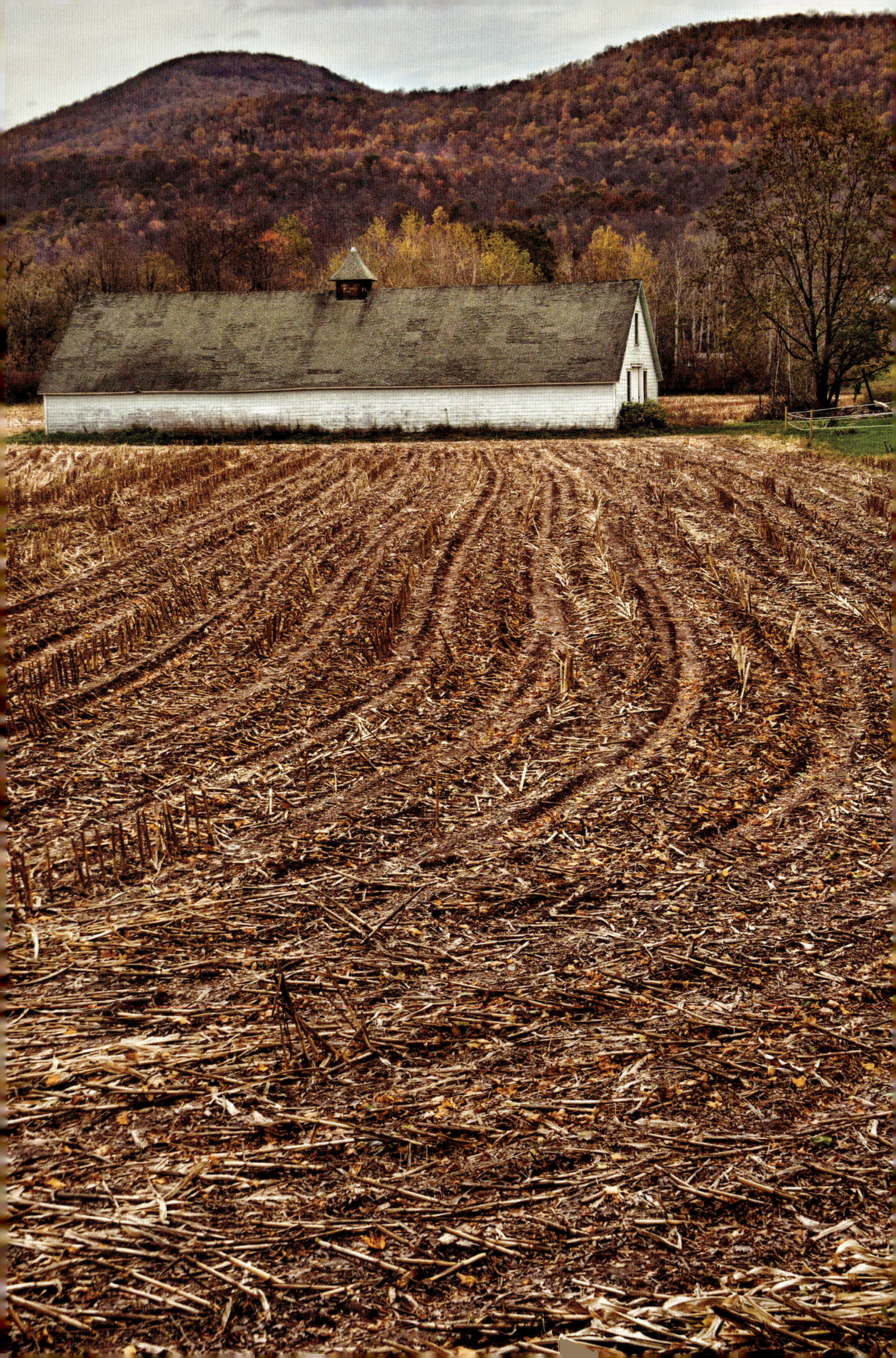

**previous page**—New York.

**right**—*Age Before Beauty.* New Hampshire.

**below**—West Virginia.

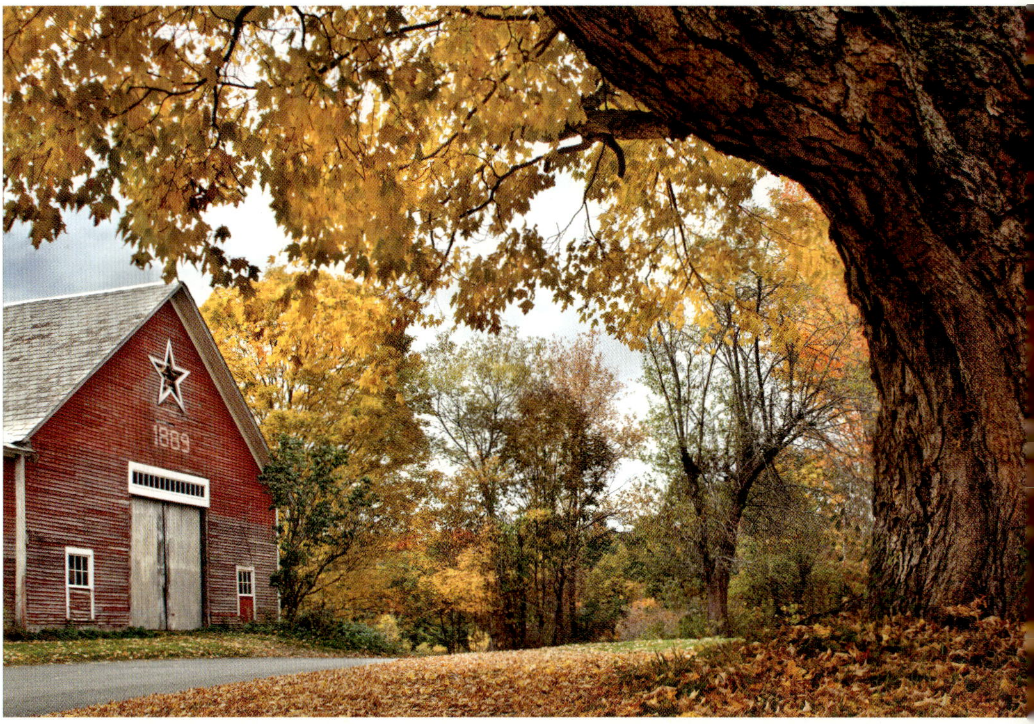


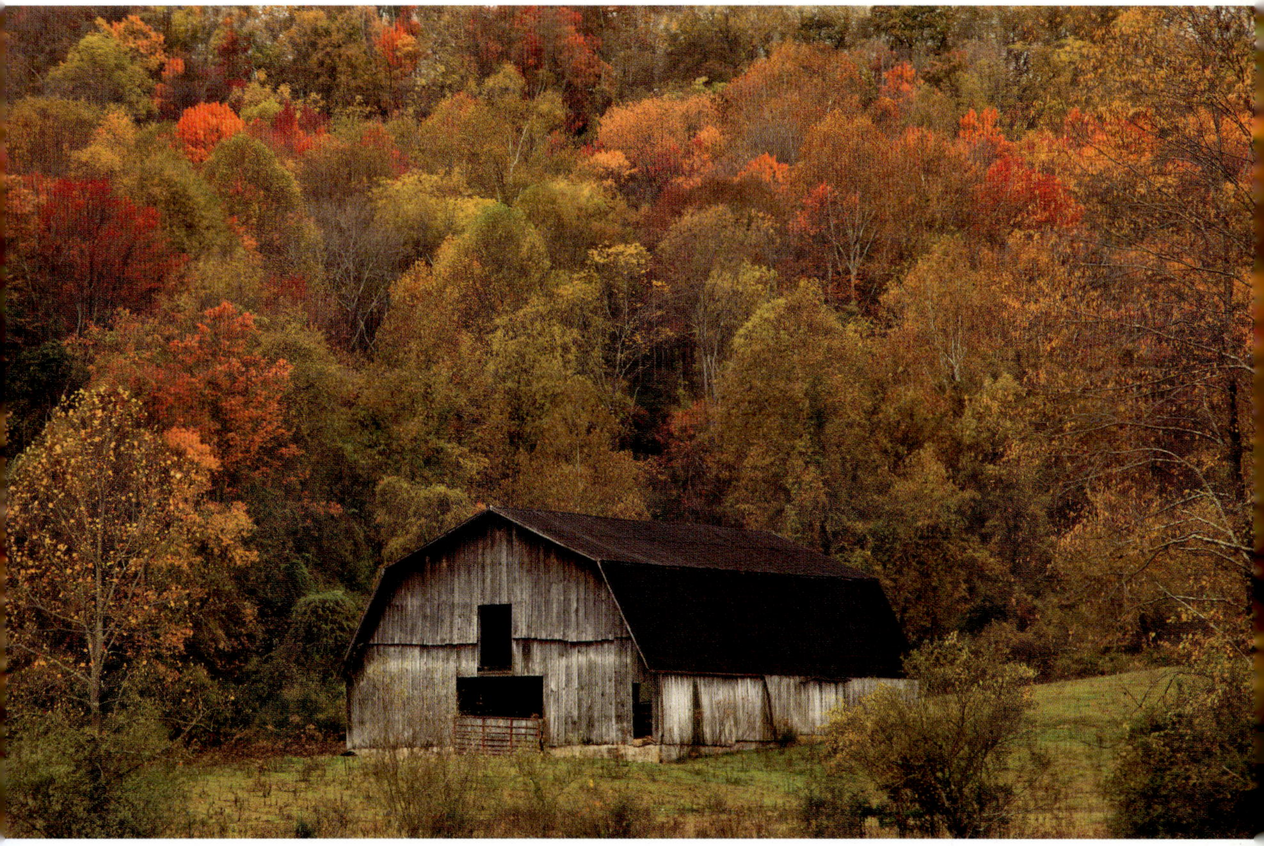

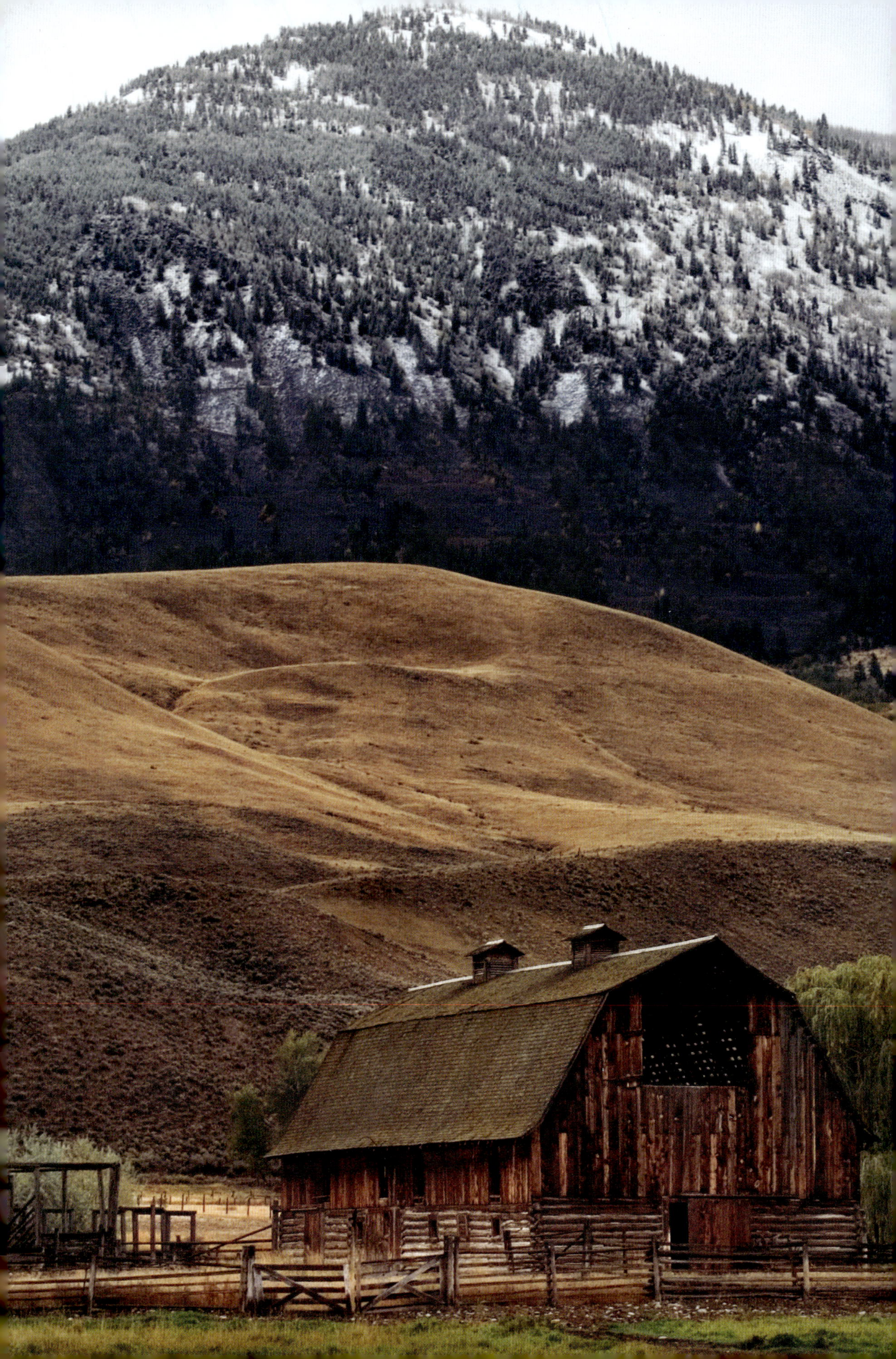

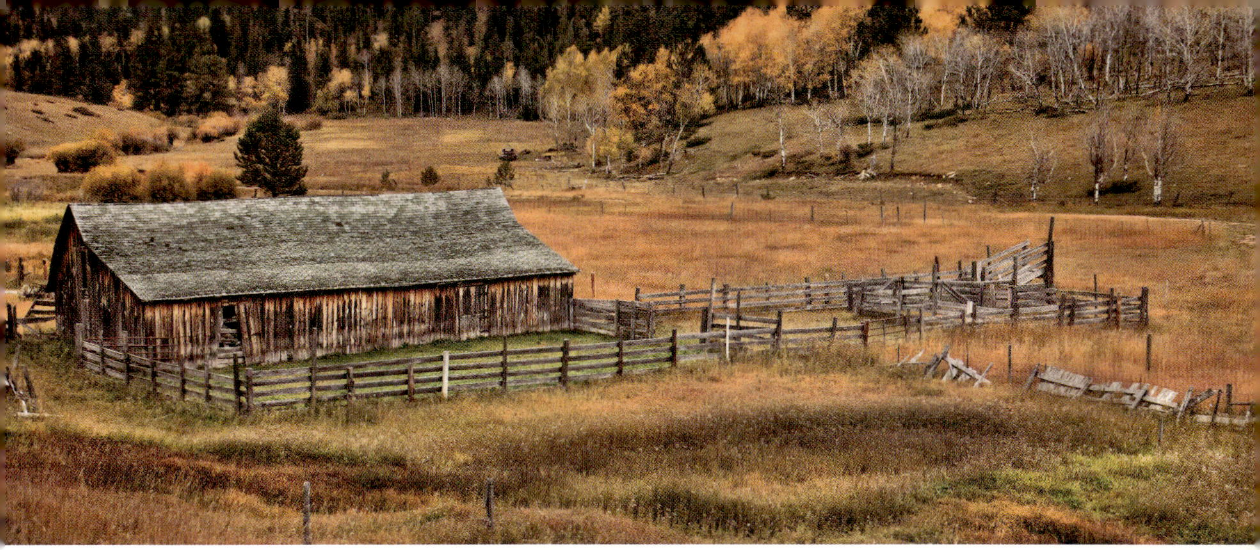

**previous page**—*First Snow*. Idaho.

**top**—Wyoming.

**center**—*Craftsmanship*. Wyoming.

**bottom**—*Abandoned Solitude*. Wyoming.

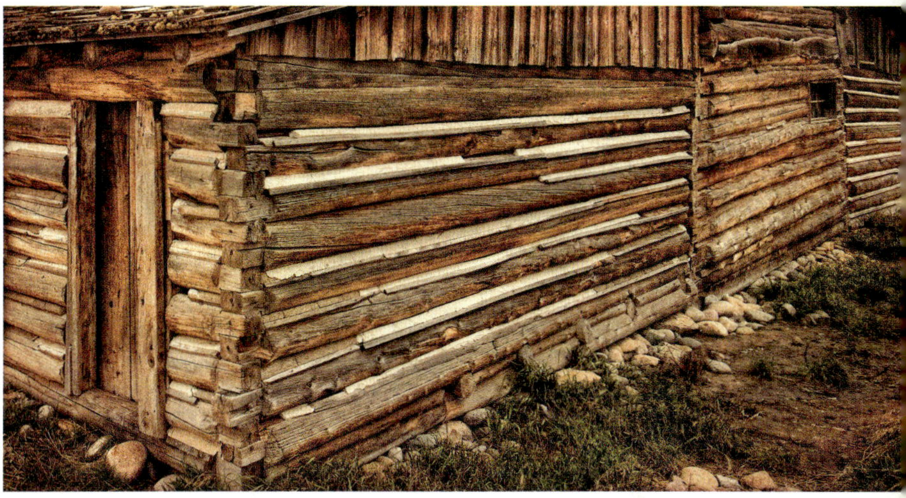

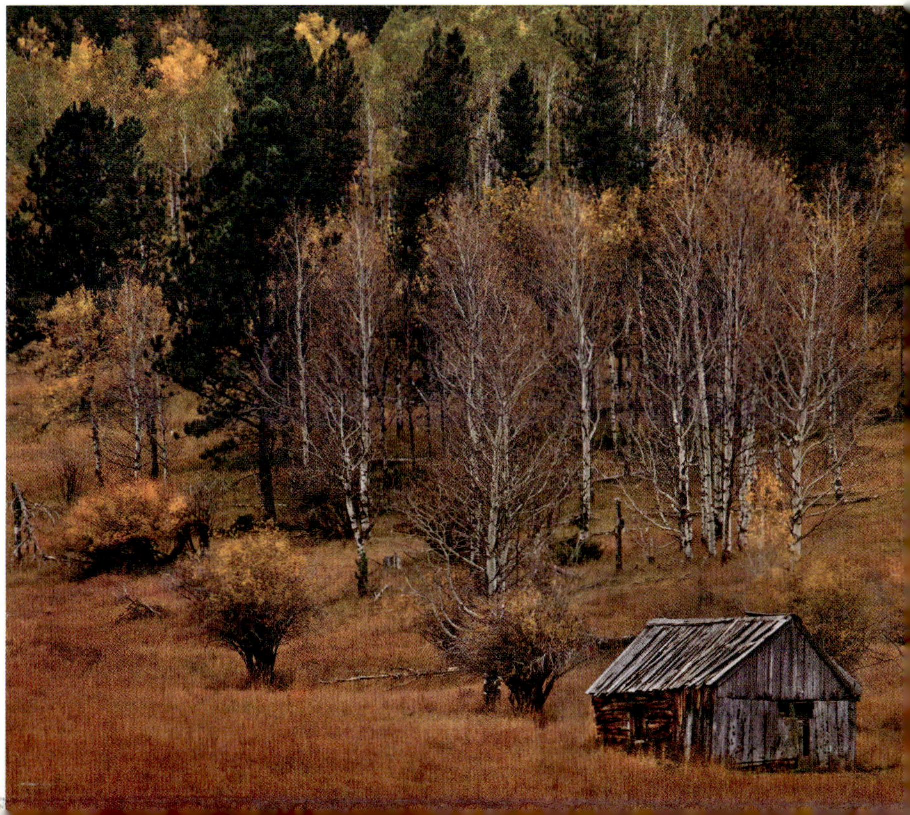

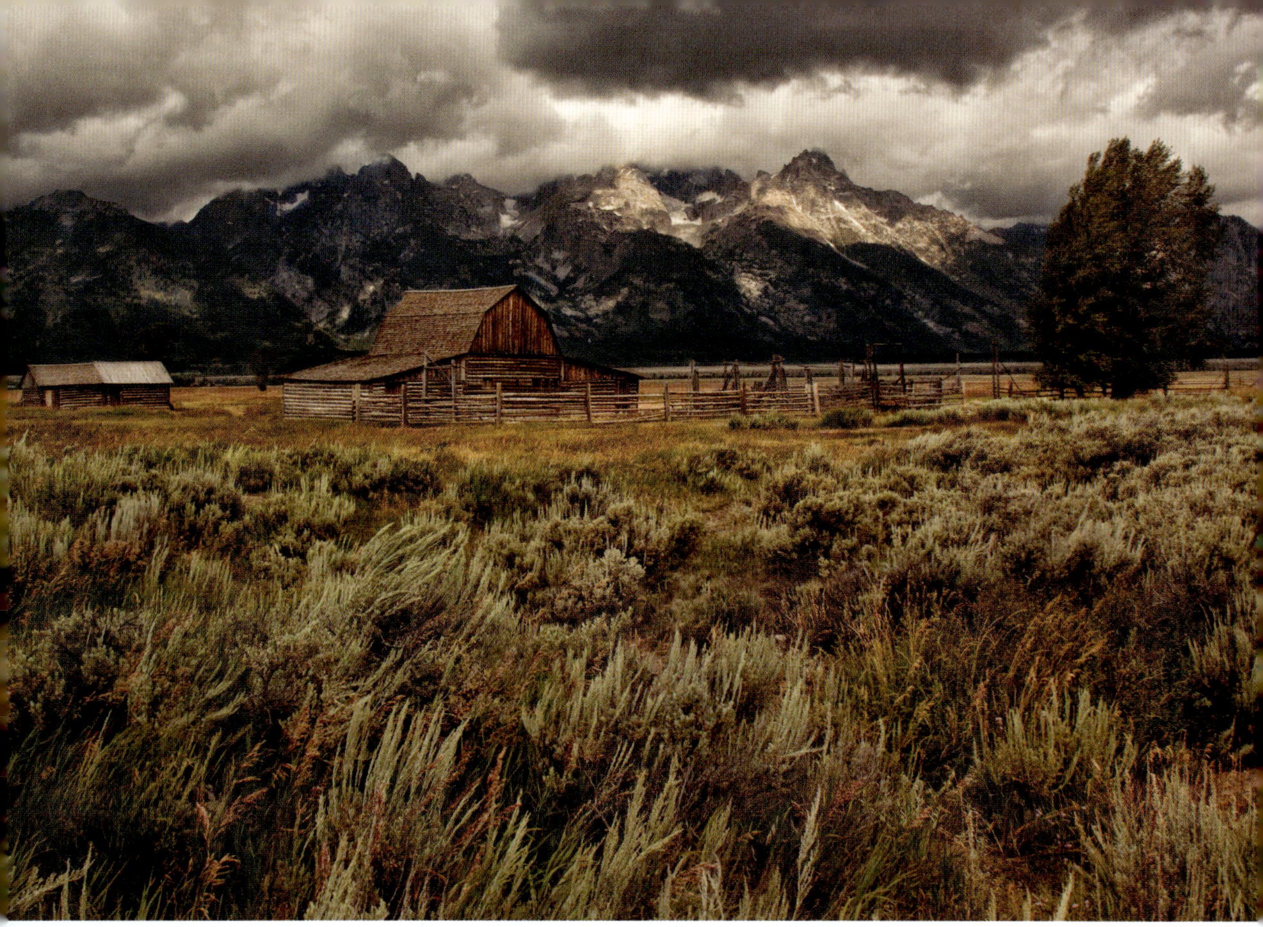

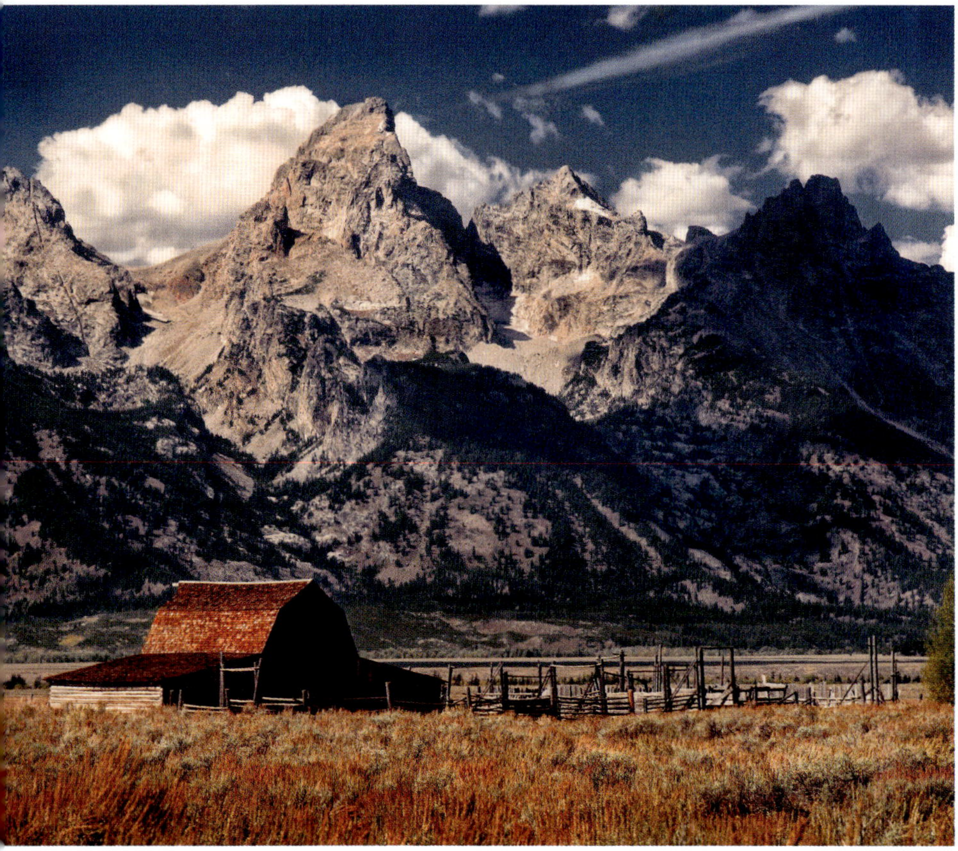

**above**—The sweet smell of mountain sage. Wyoming.

**left**—*Homestead in the Tetons.* Photographed on film. This Wyoming barn, built by a homesteader, is the most-photographed barn in America.

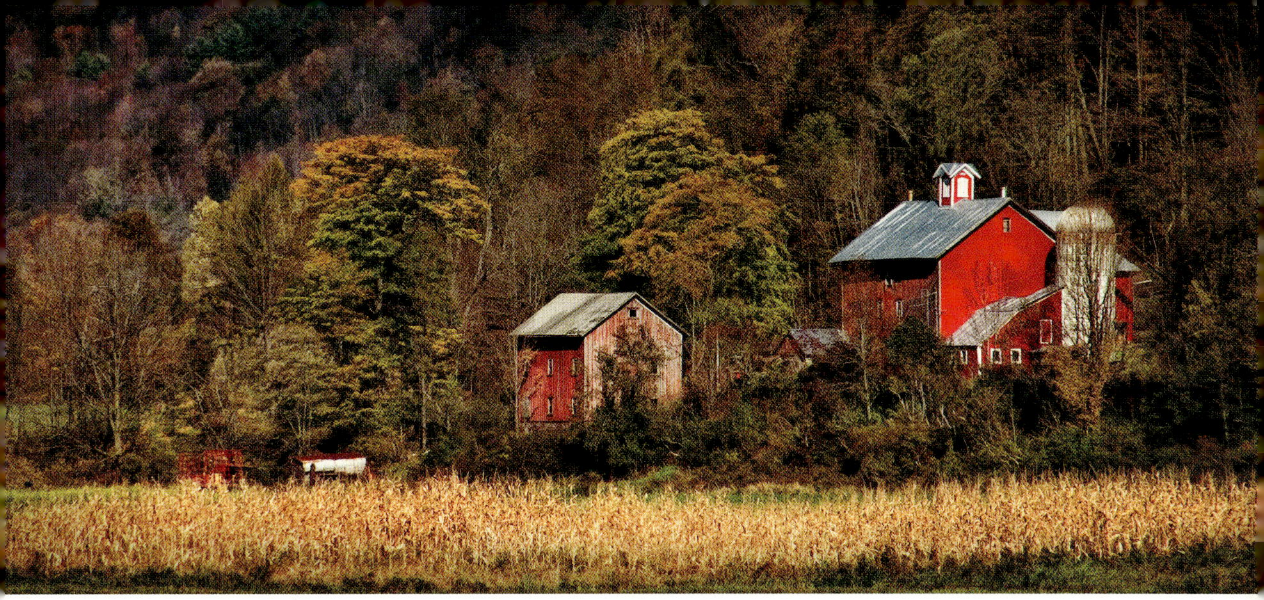

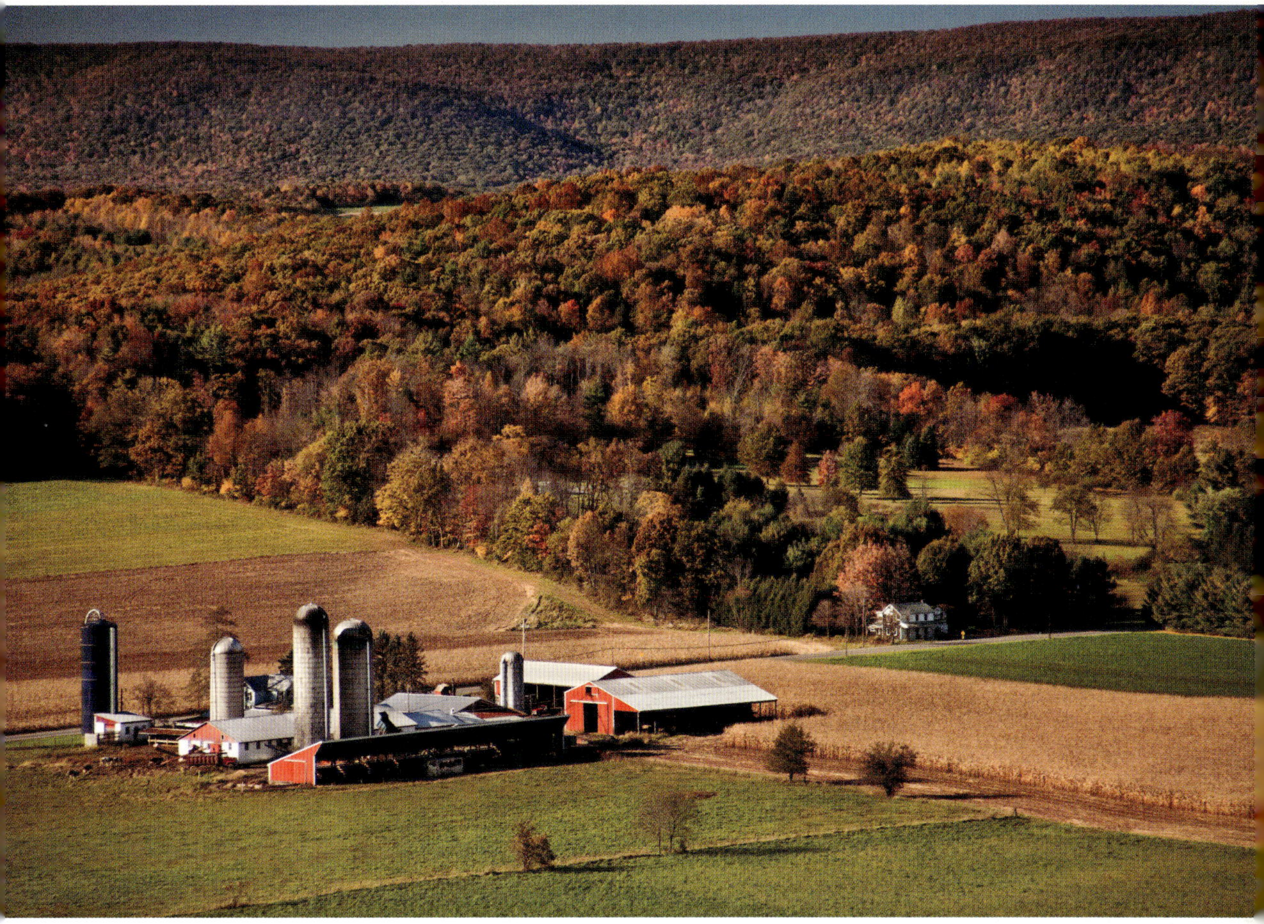

**top**—New York.

**bottom**—Pennsylvania.

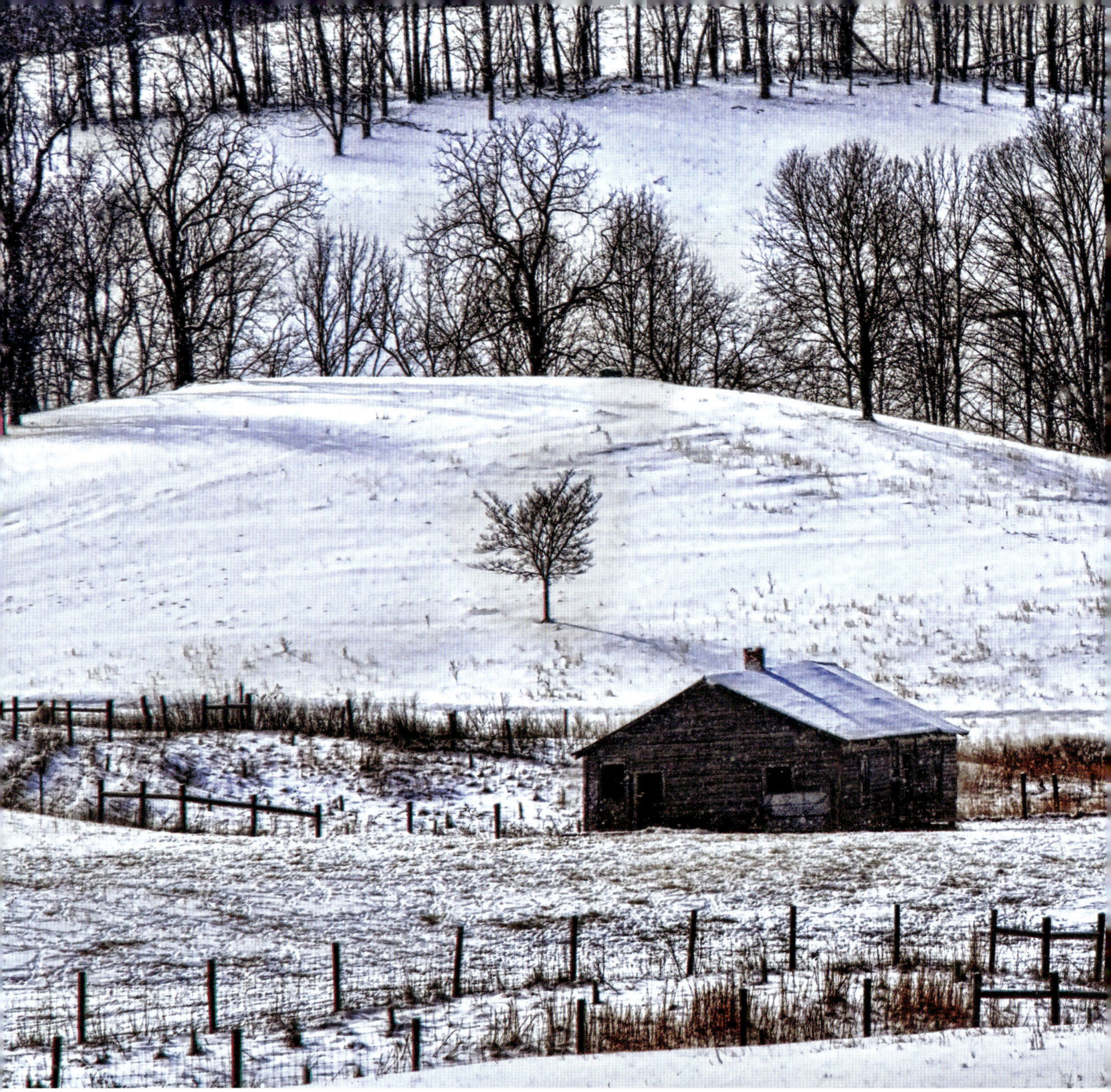

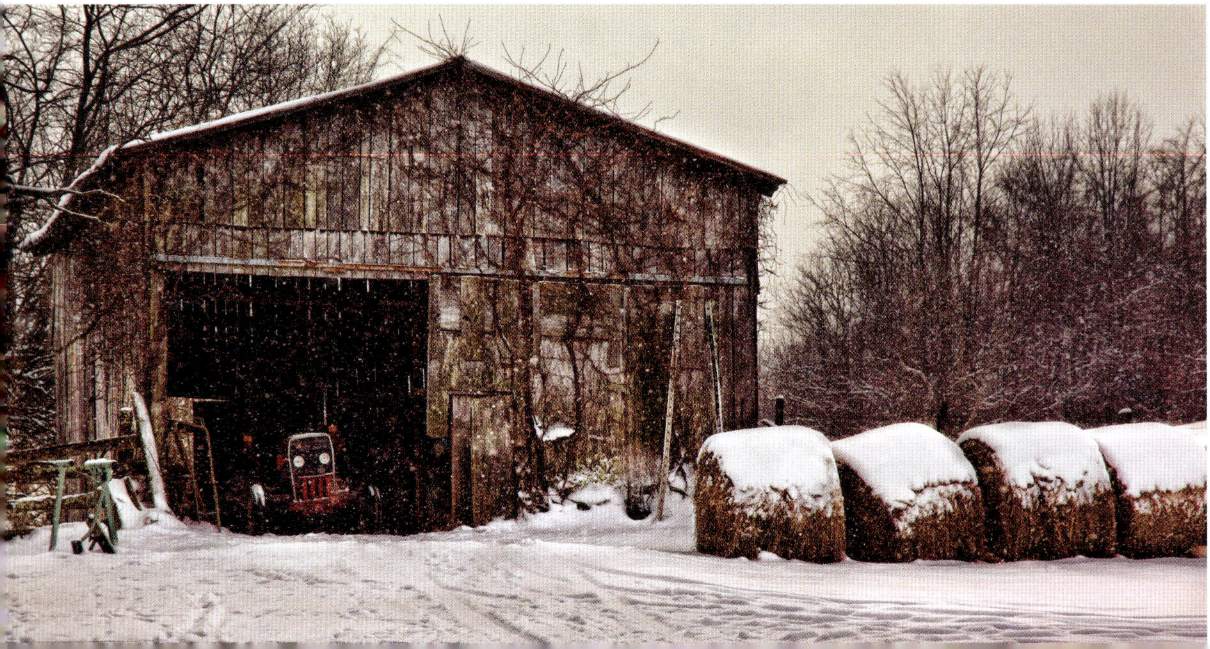

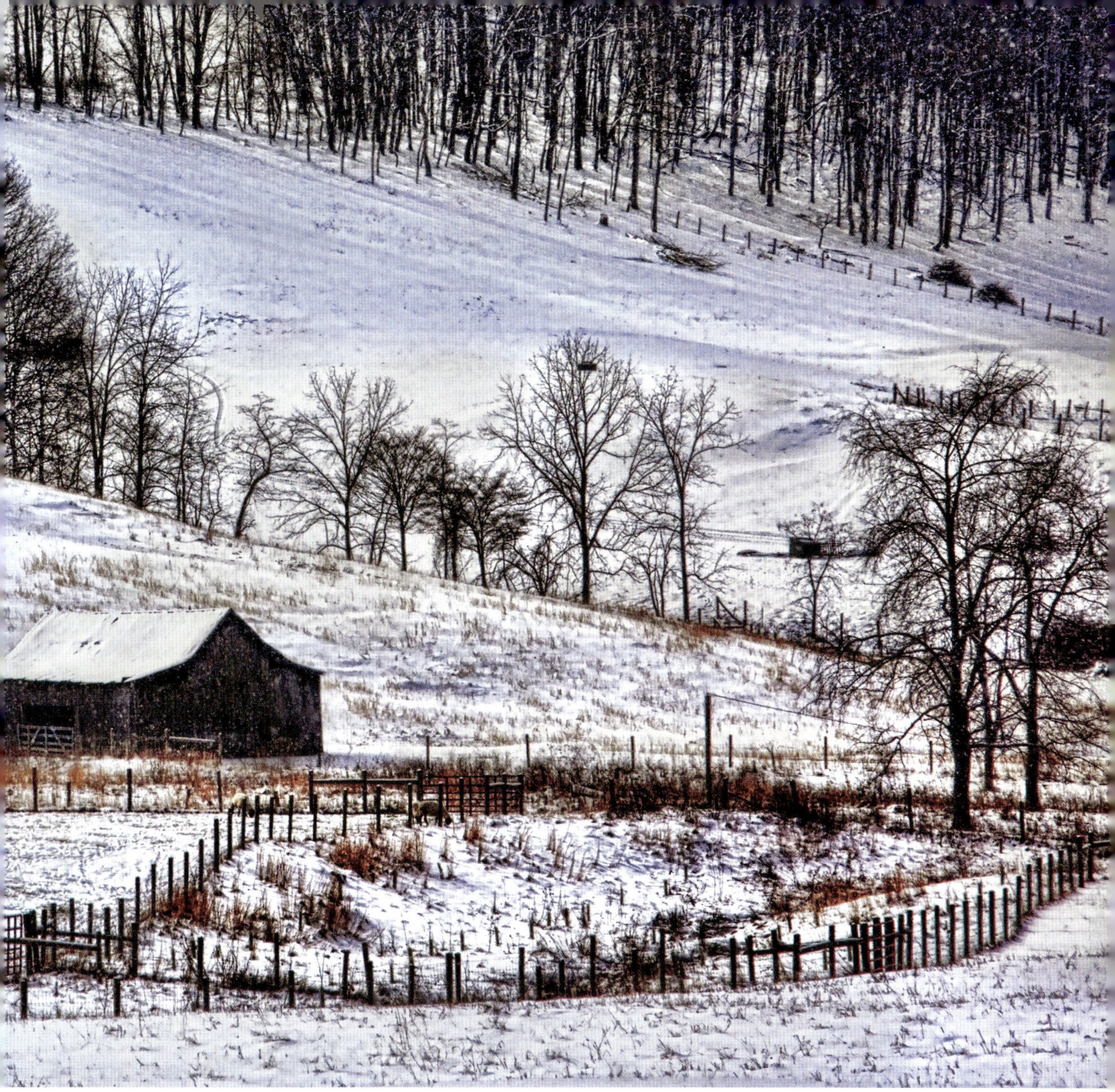

above—*Winter Solitude*. Virginia.

previous page, bottom—Virginia.

right—Winter grazing. Kentucky.

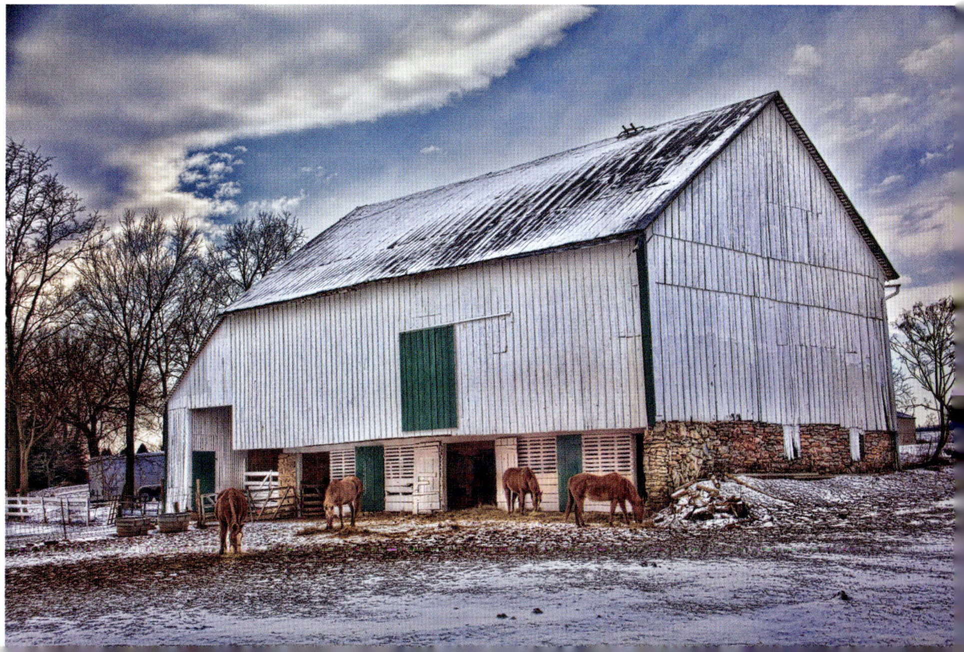

**right, top and bottom**—*Abandoned Dreams*. Kentucky.

**below**—Pennsylvania.

**bottom**—Vermont.

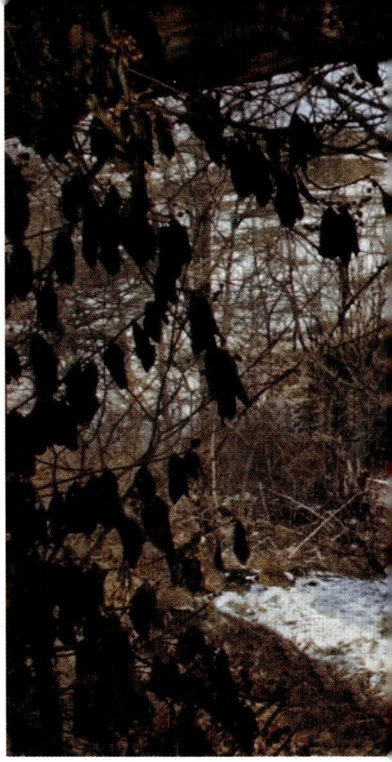

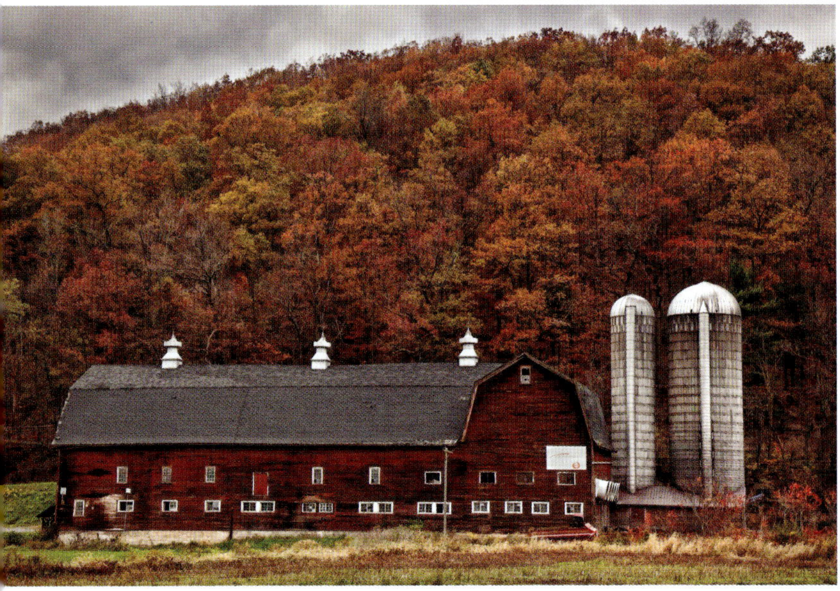

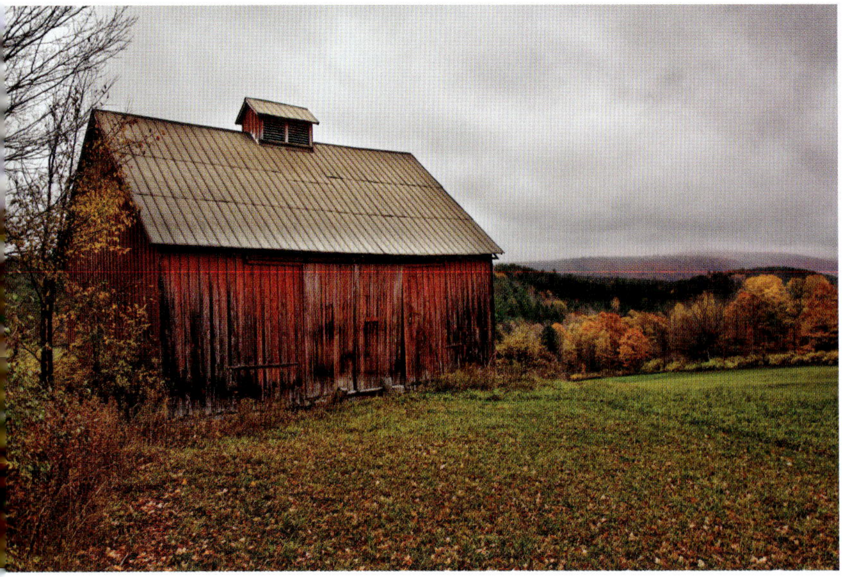

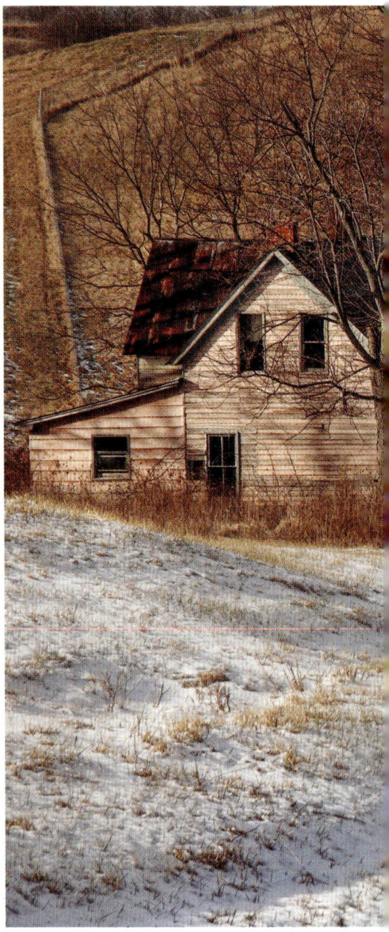

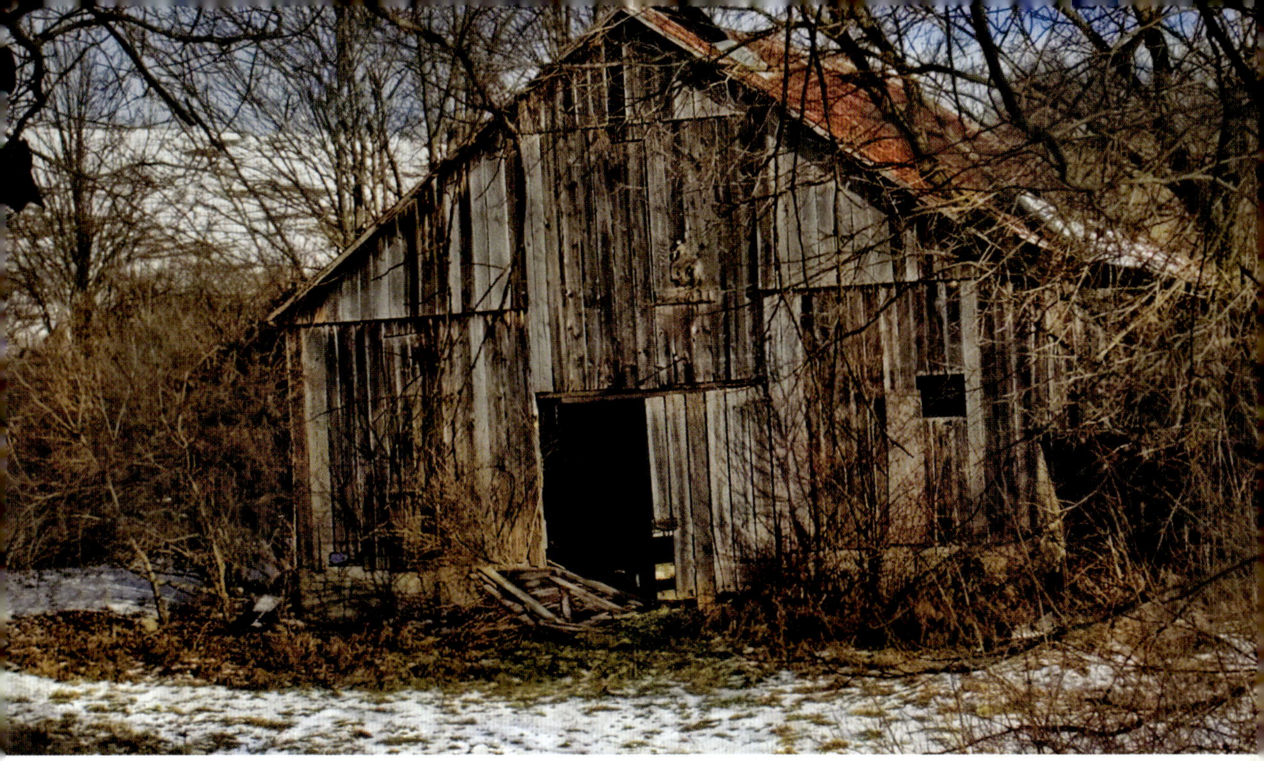

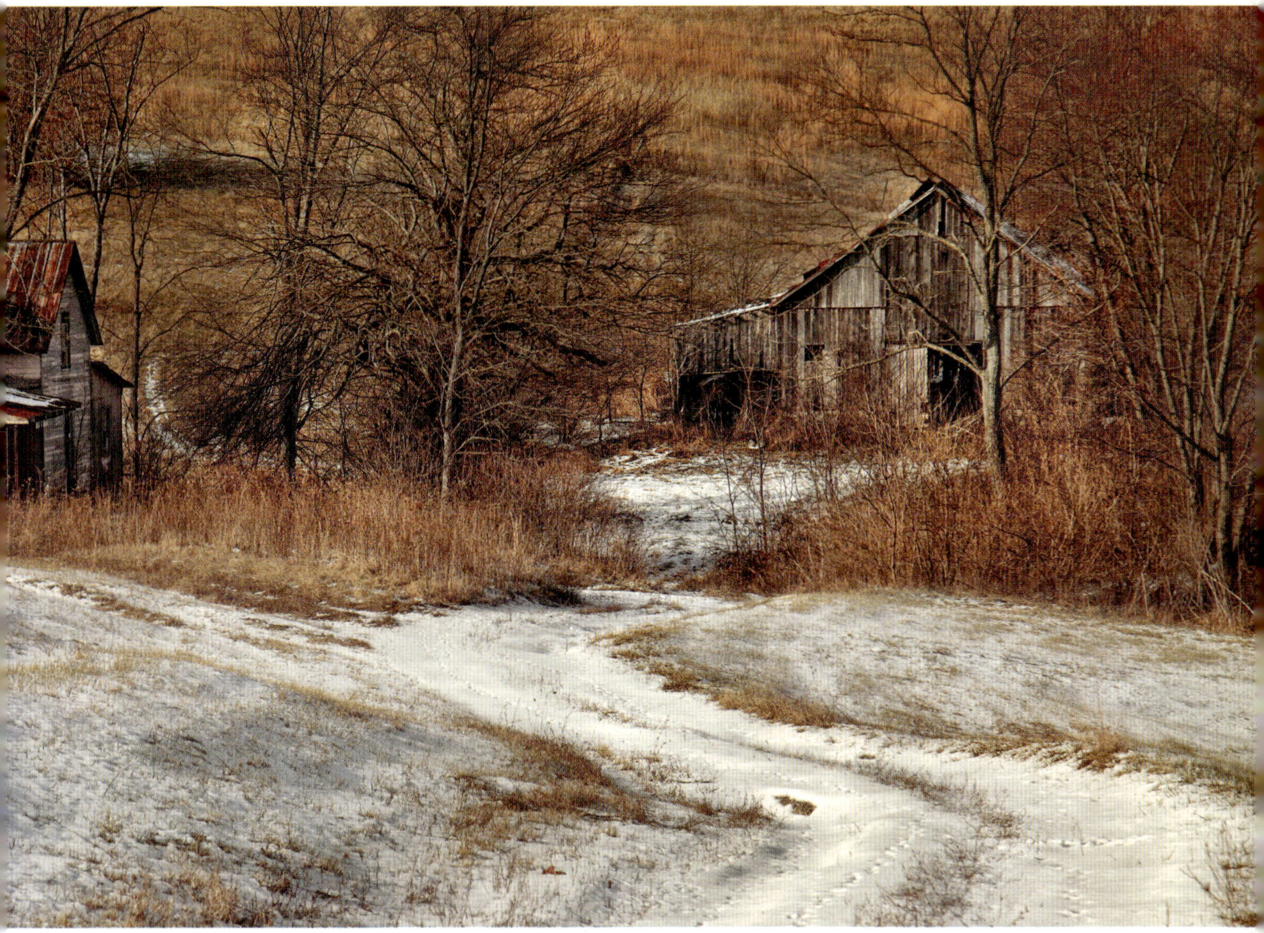

# Composition

A single barn can have many faces, depending on the angle from which it is viewed and photographed. Even the light, season, and time of day can have a powerful impact on the appearance of the subject and scene.

This chapter includes images that show this tendency. As you'll see, the images, arranged in a group, can make for a nice photographic presentation that helps to tell a story.

**"A good photograph is knowing where to stand."**

**—Ansel Adams**

**below**—Gordon, Nebraska.

**top and bottom, following page**—Two different perspectives of the same barn. Gordon, Nebraska.

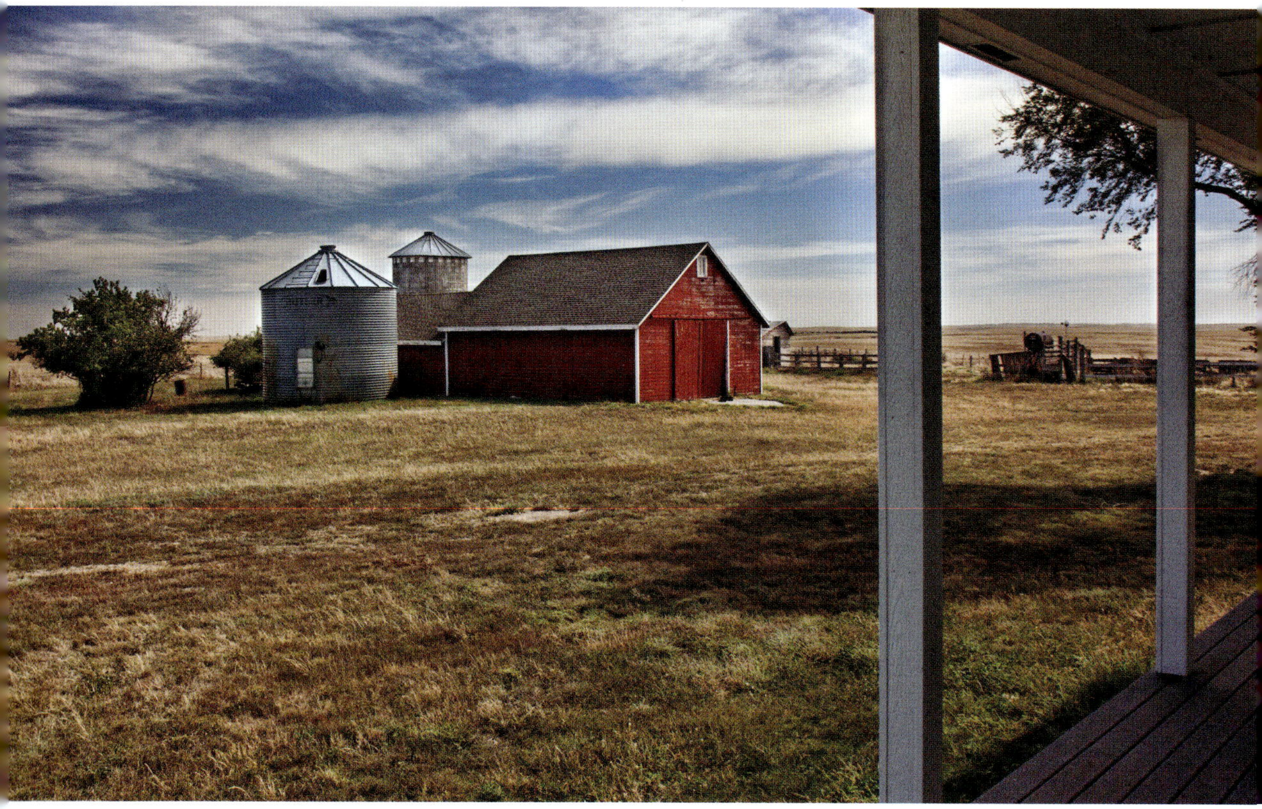

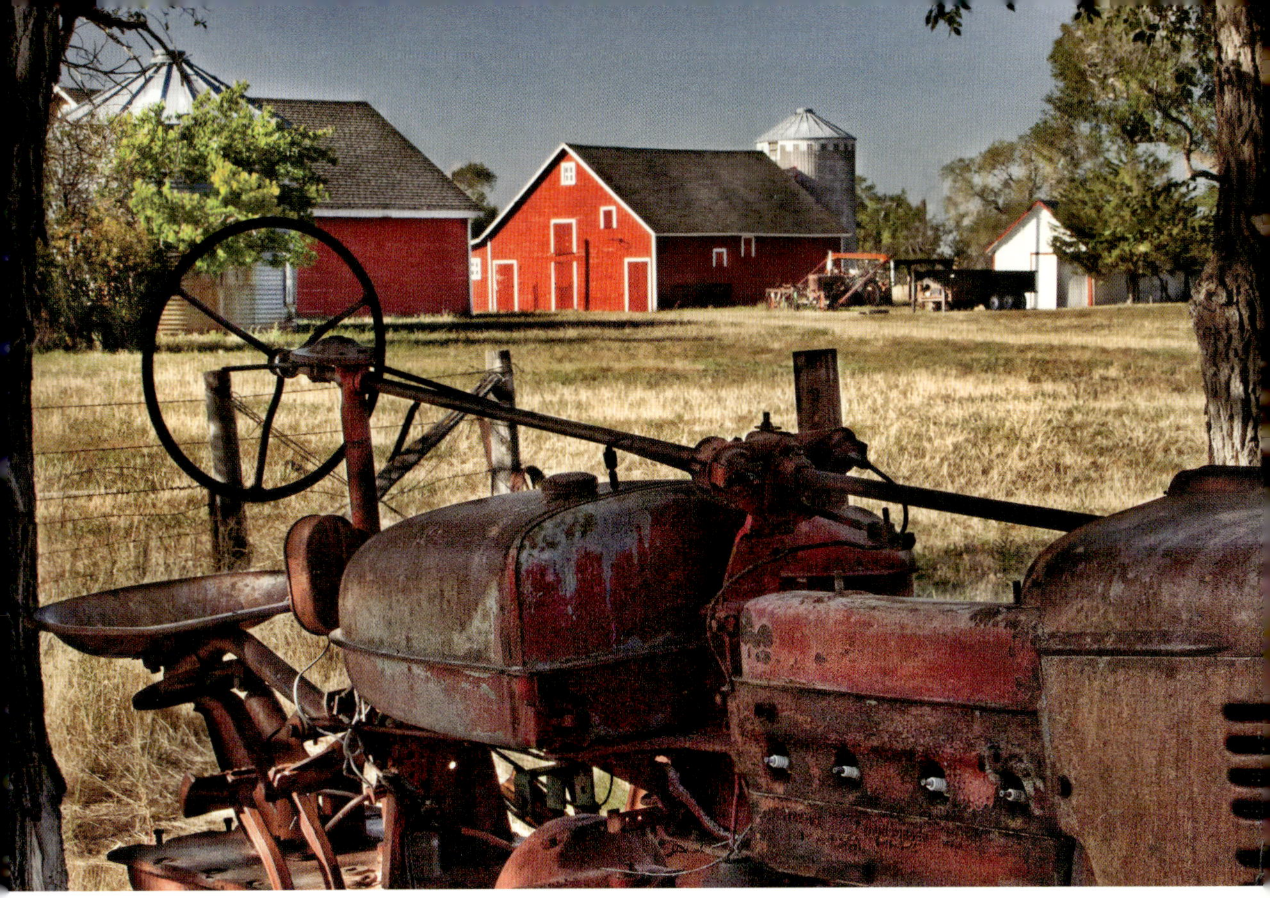

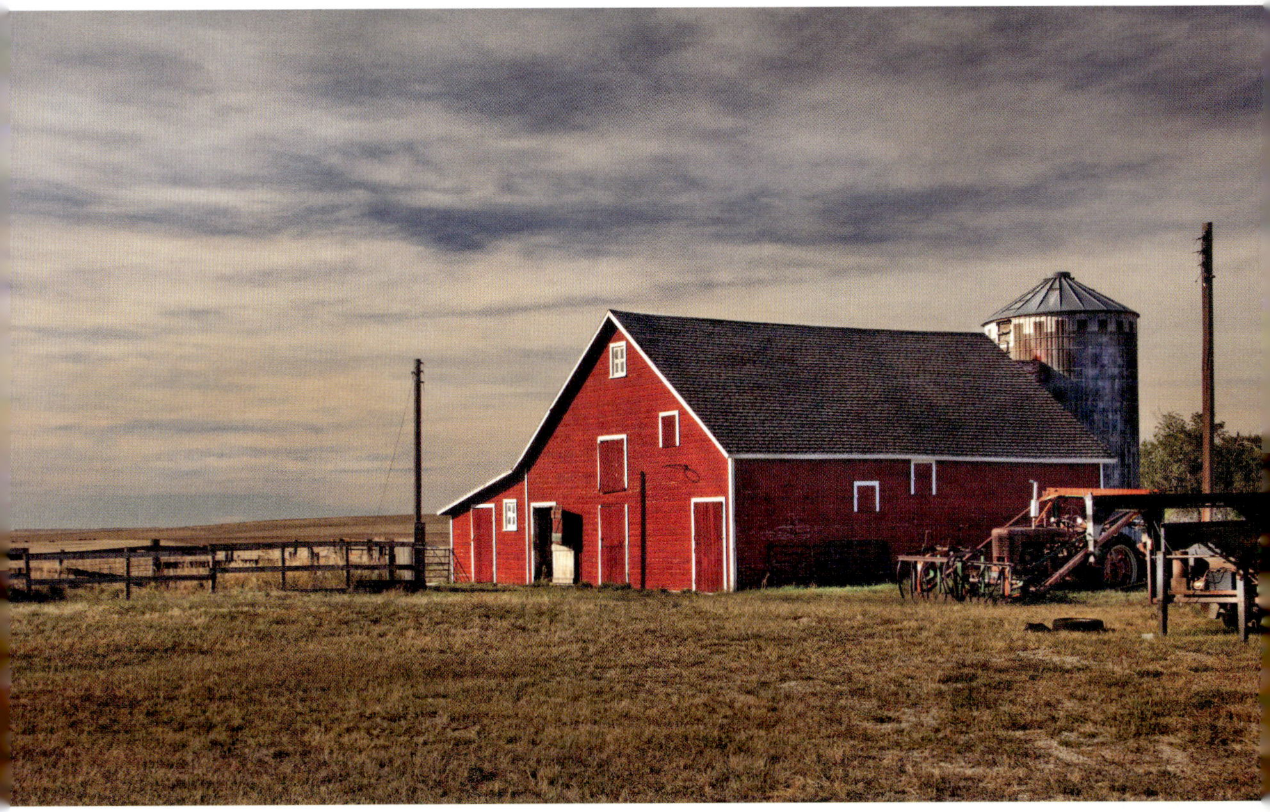

Golden hour occurs twice a day: during the hour after sunrise and the hour before sunset. During this time, the light takes on a warm, golden hue that can enhance the beauty of any subject.

It's not every day that I find a barn I can photograph at sunrise and sunset. When I can, the results speak for themselves.

**below**—Golden hour. Nebraska.

**following page**—The front of a barn, photographed at sunrise, and the back photographed at sunset. Gordon, Nebraska.

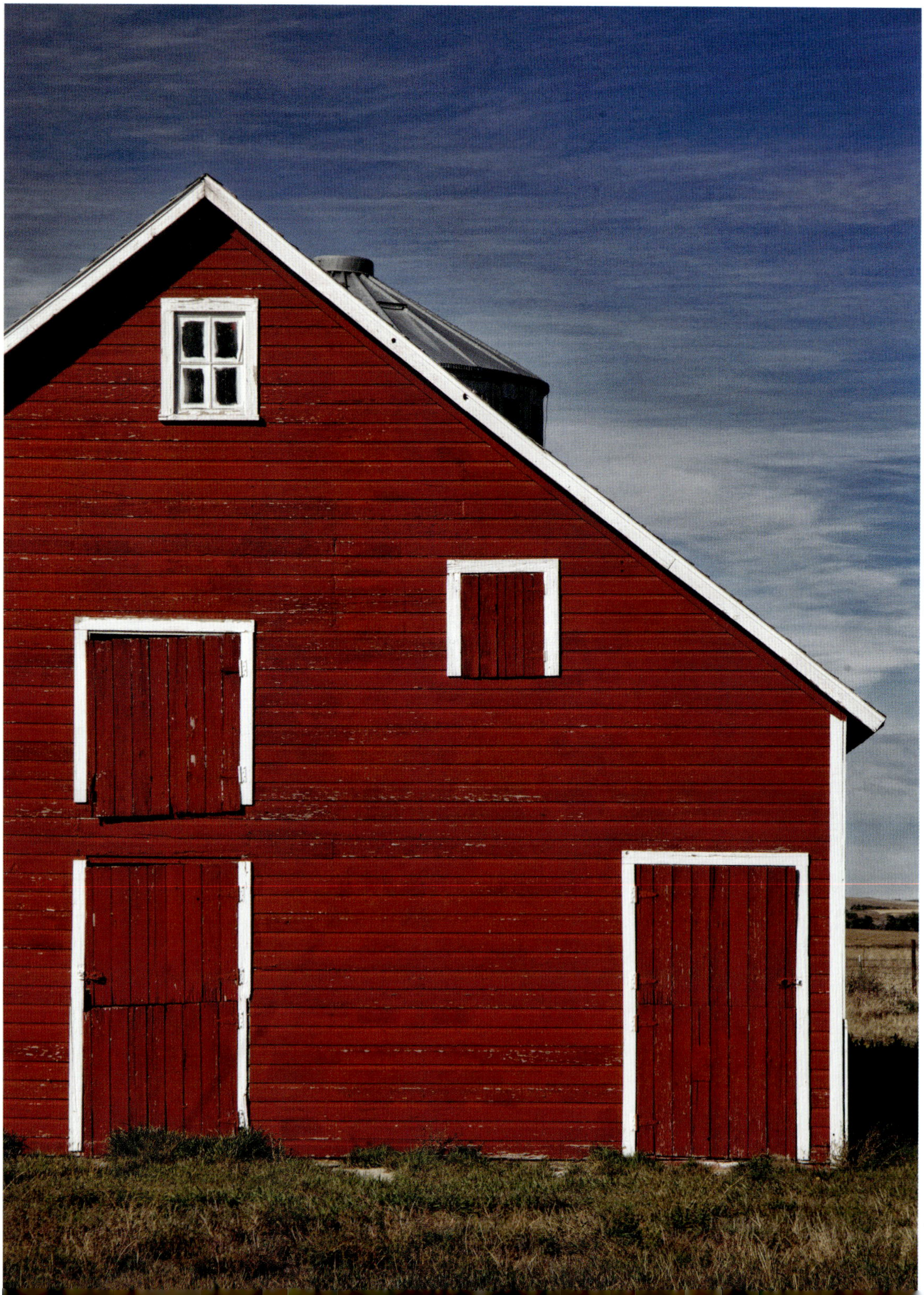

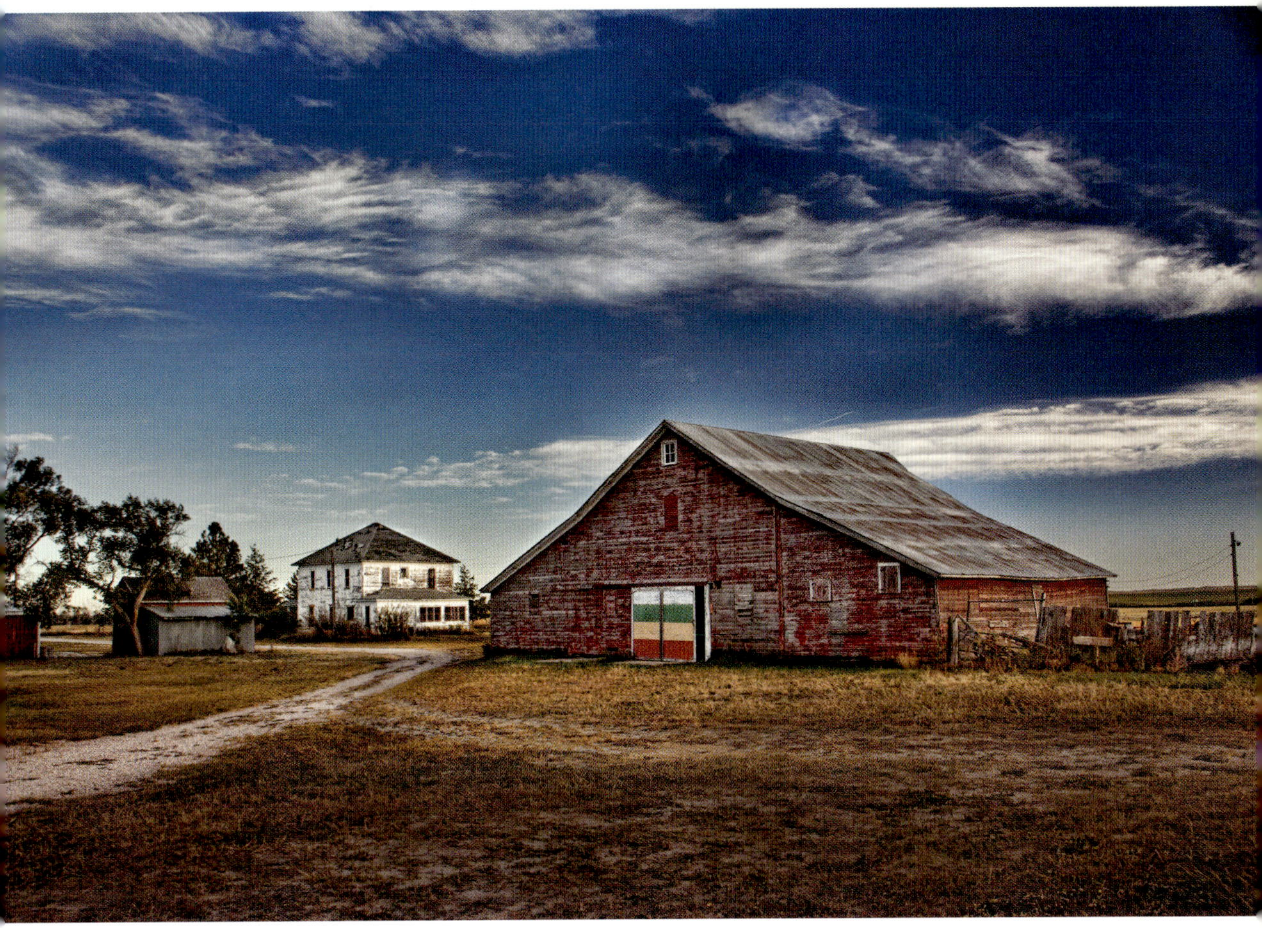

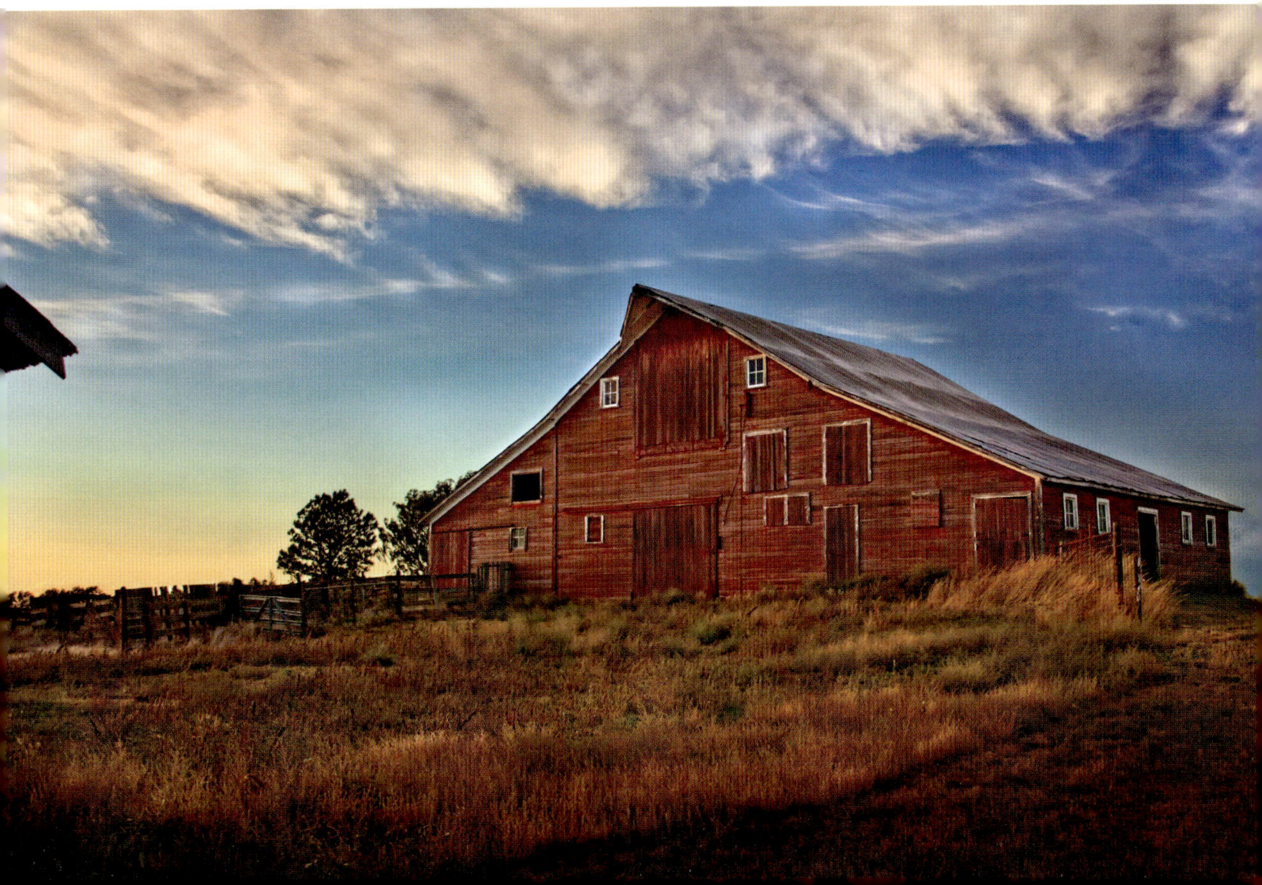

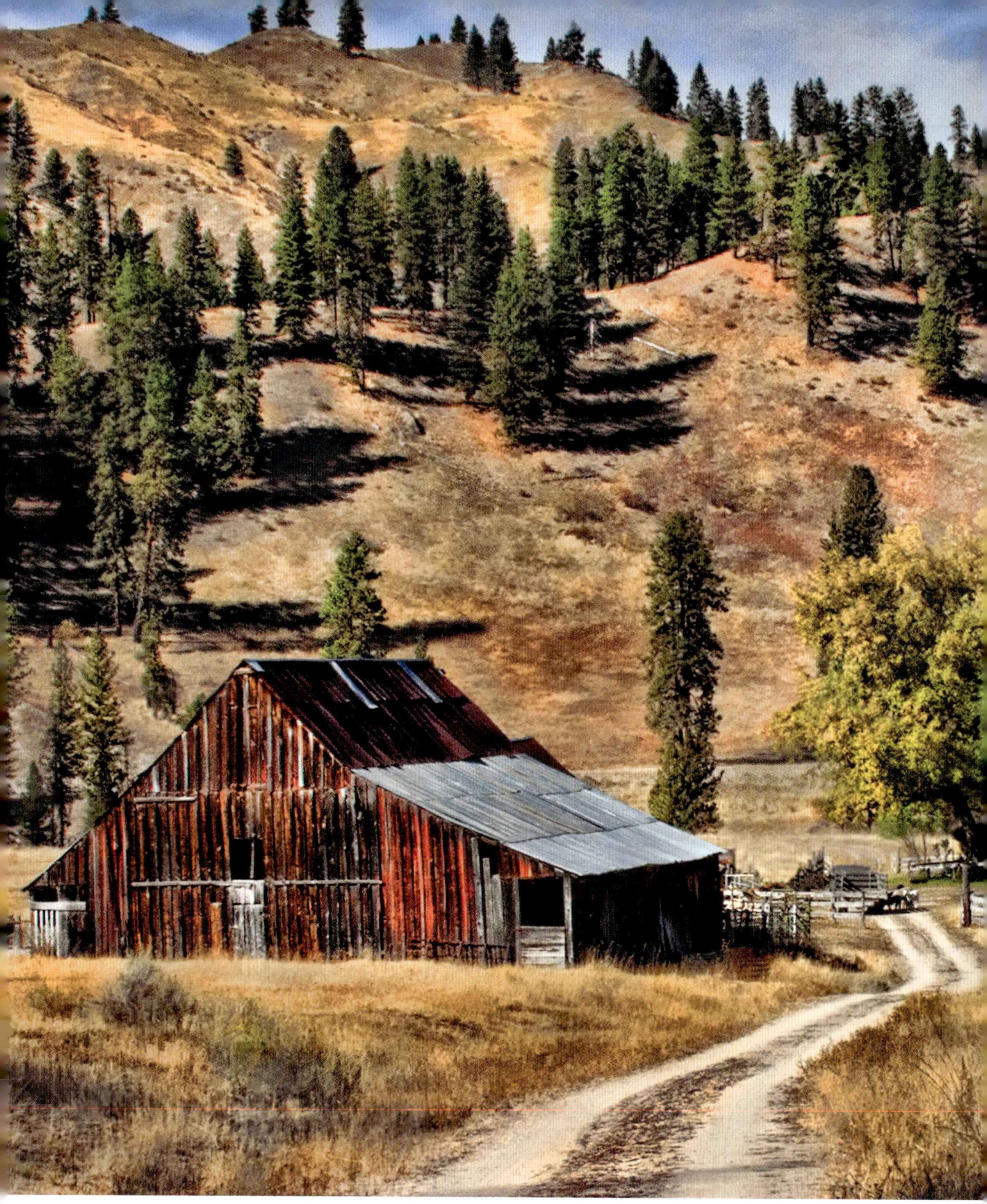

**above**—*Alone in the Foothills.* Idaho.

**following page**—Great color contrast and a unique perspective make this shot.

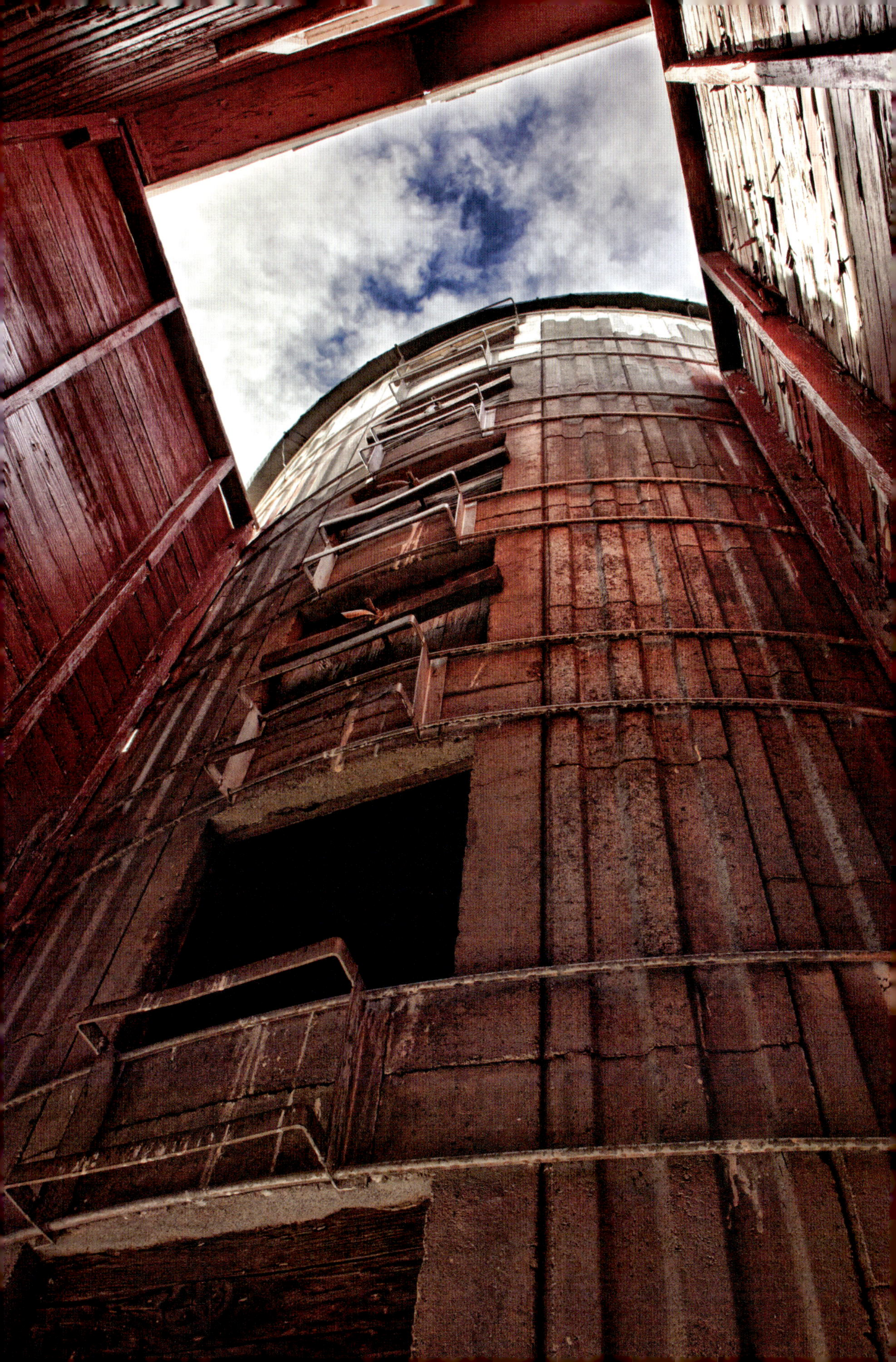

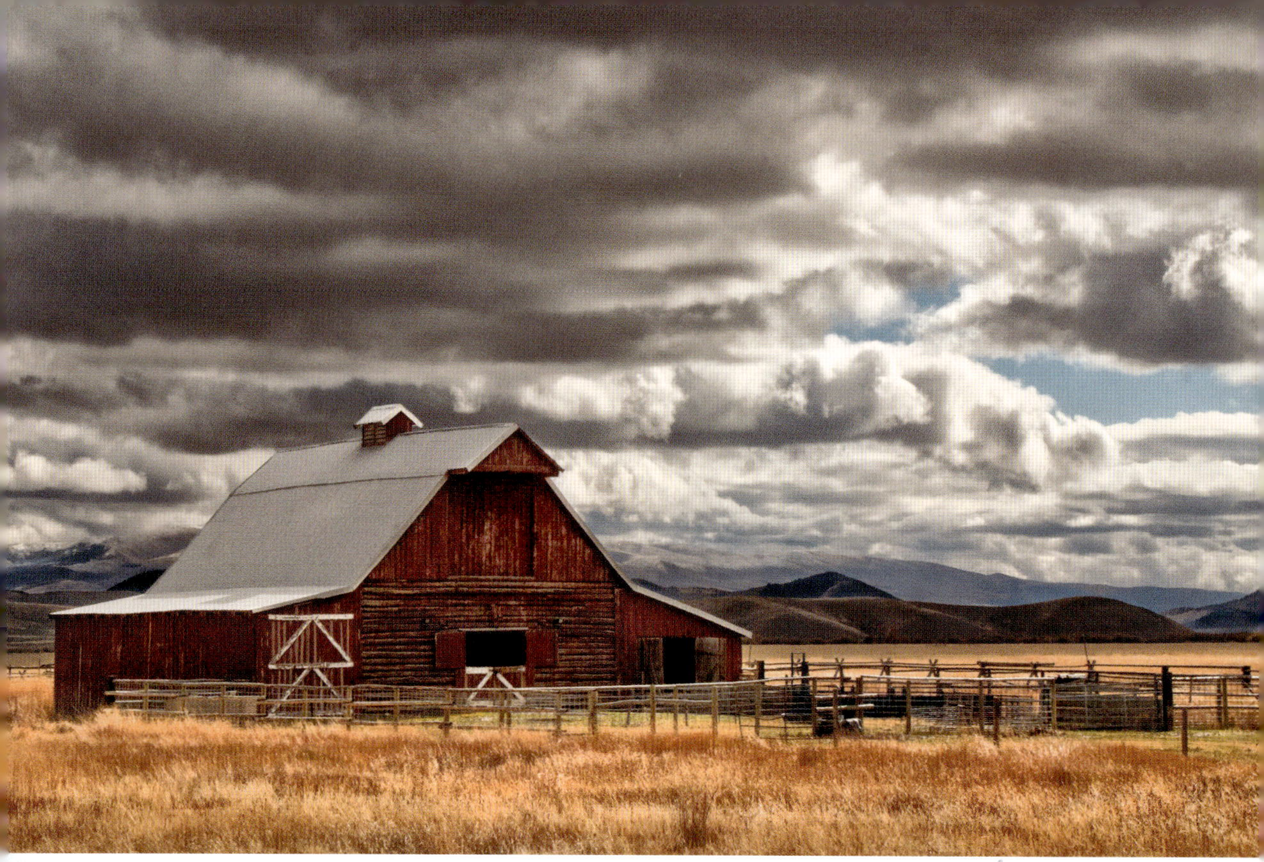

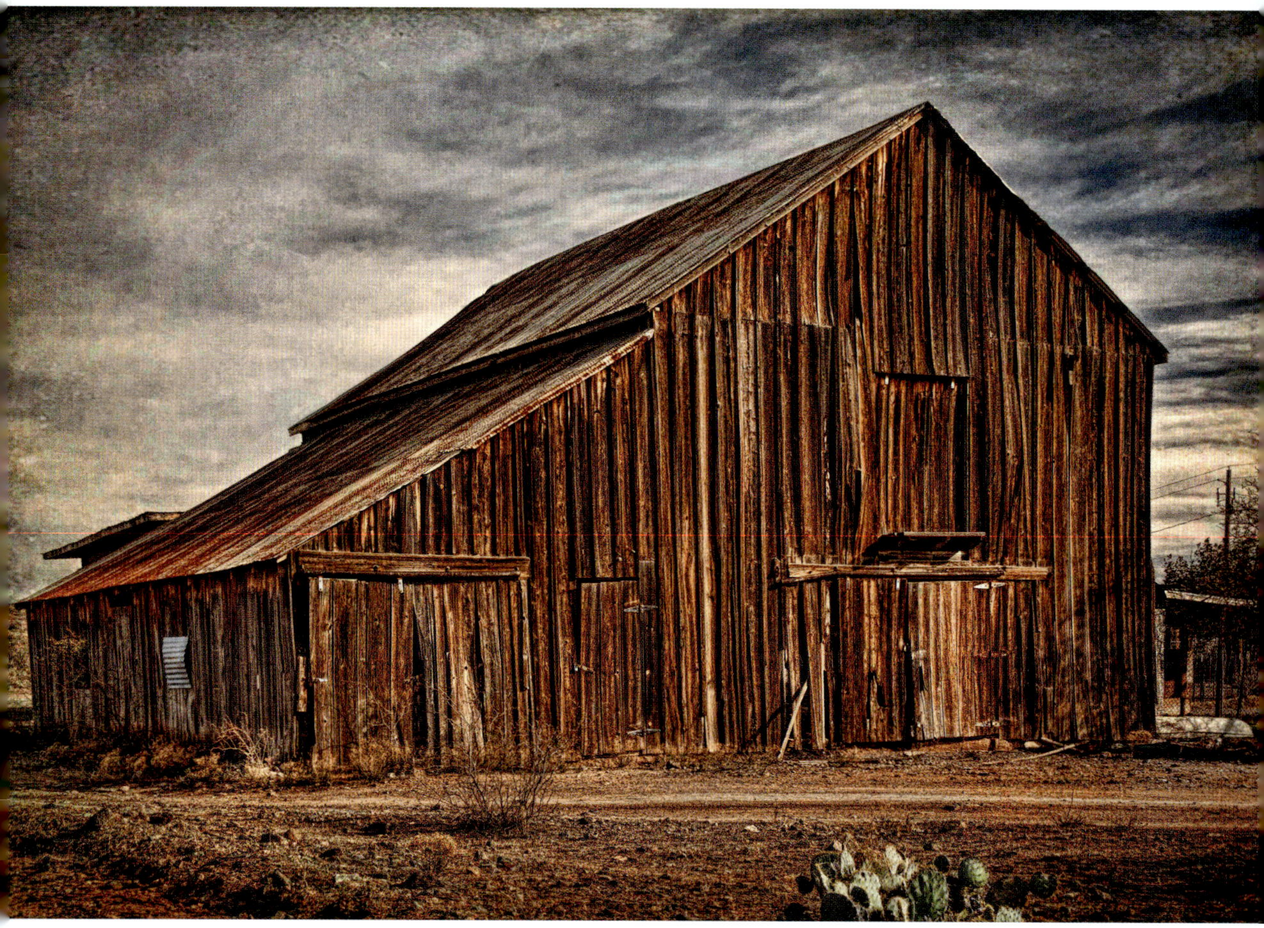

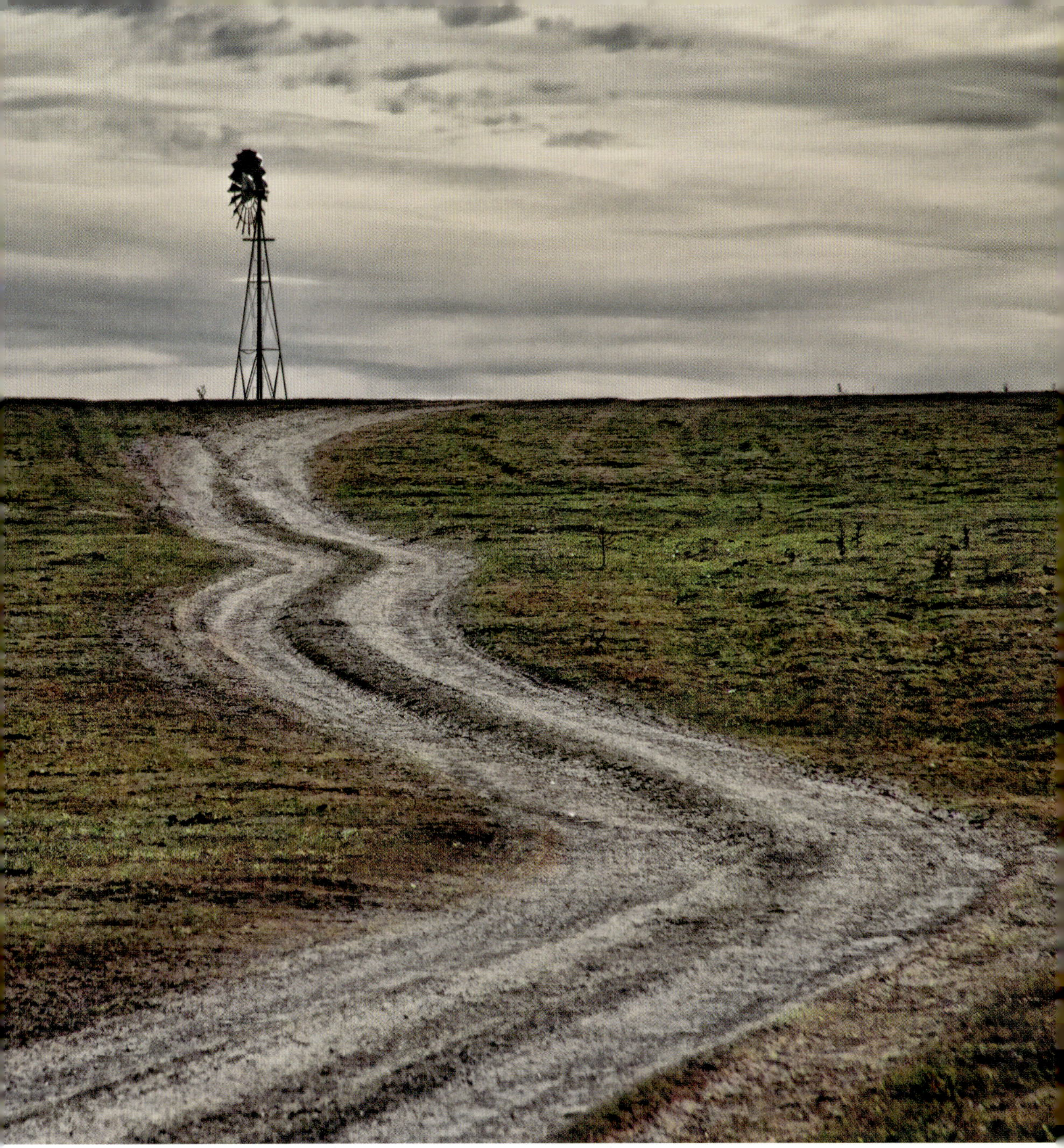

**previous page, top—**A dramatic sky frames this weathered barn in Montana.

**previous page, bottom—**Cordes Station barn. Arizona.

**above—**The great open plains of Nebraska.

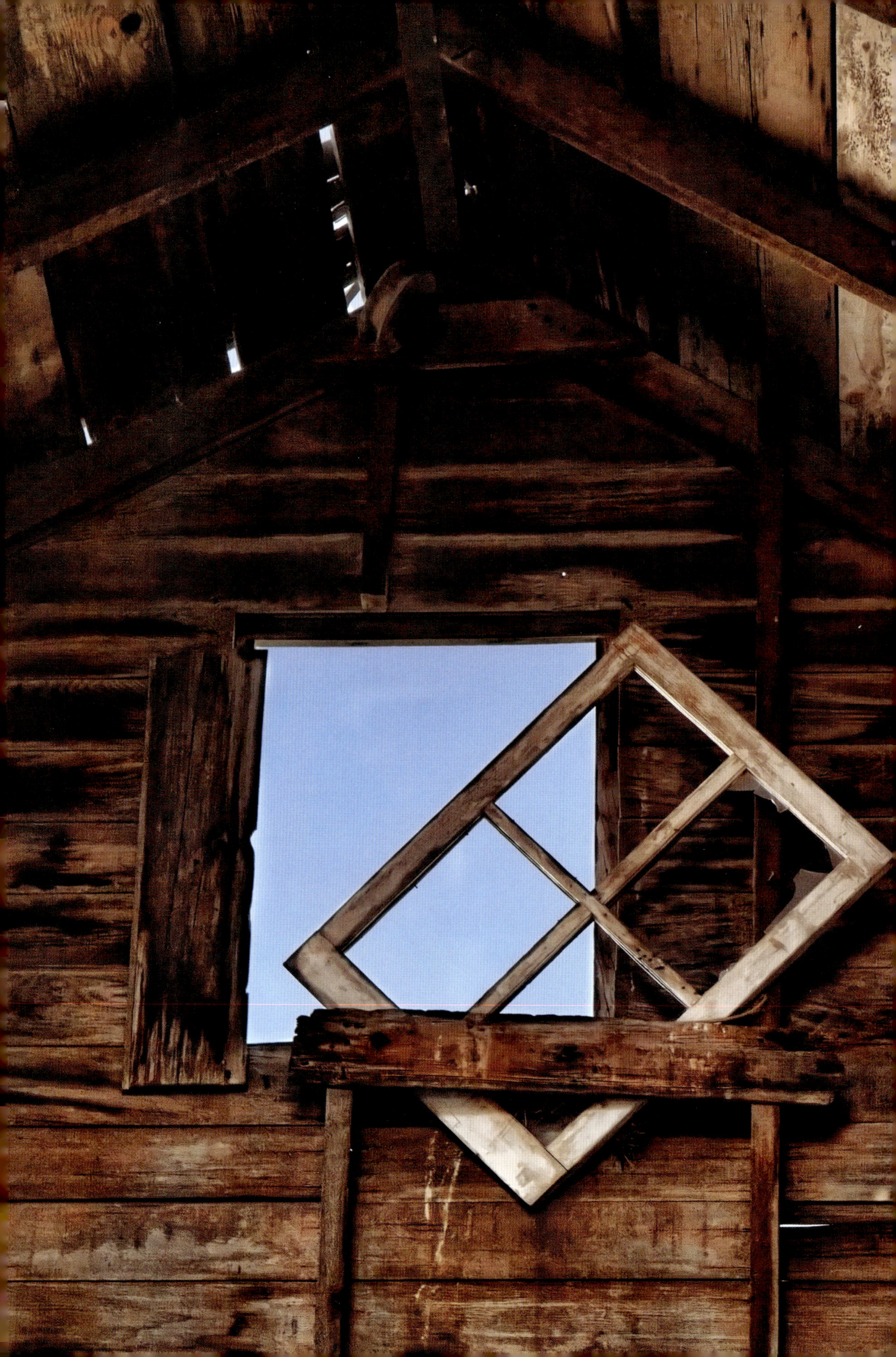

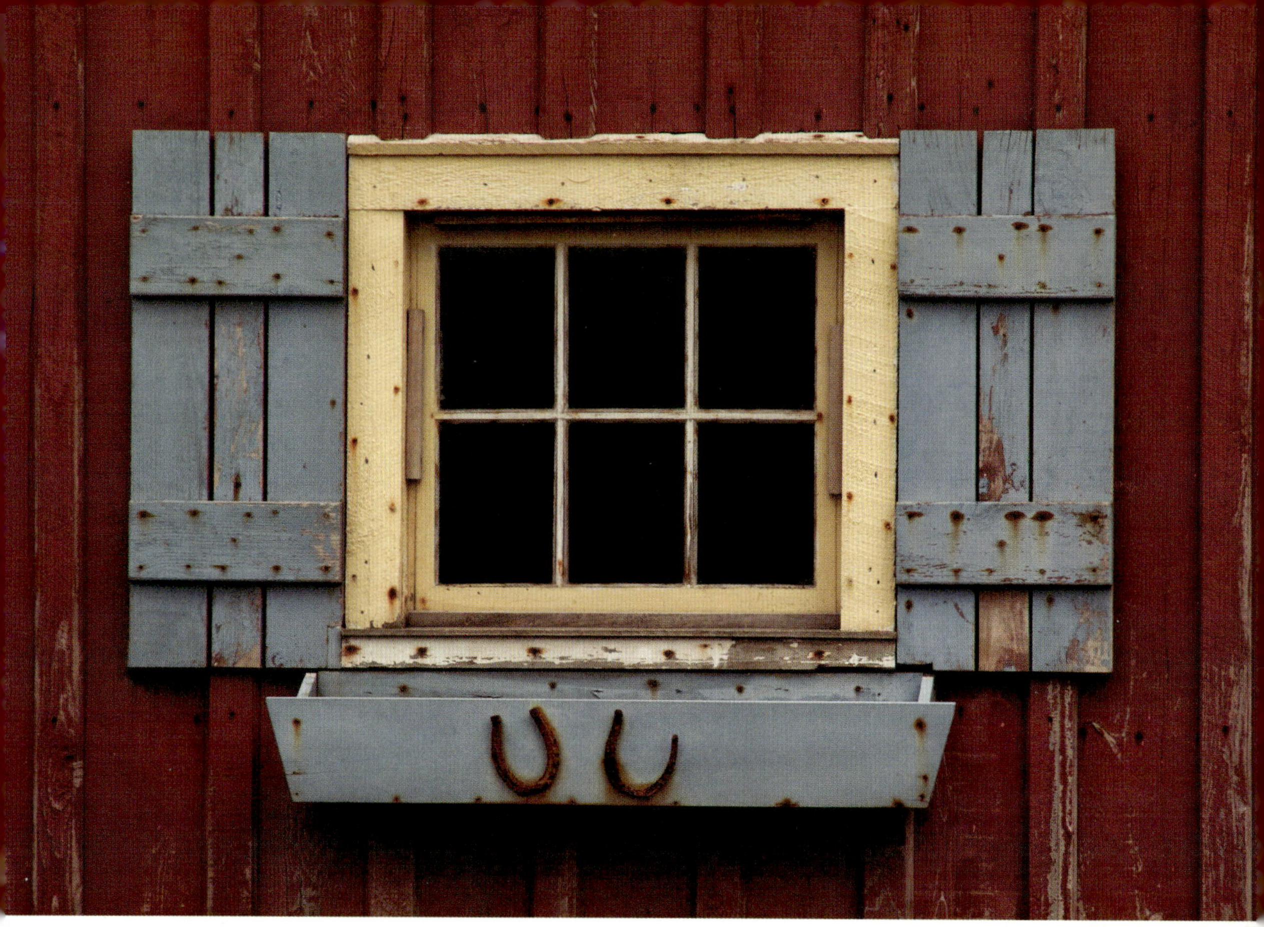

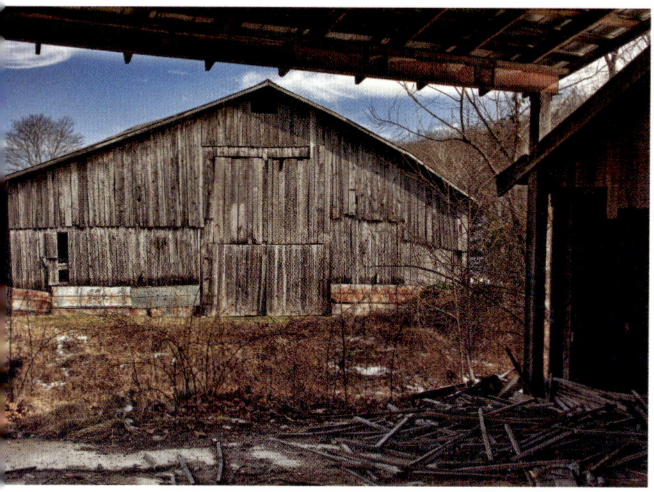

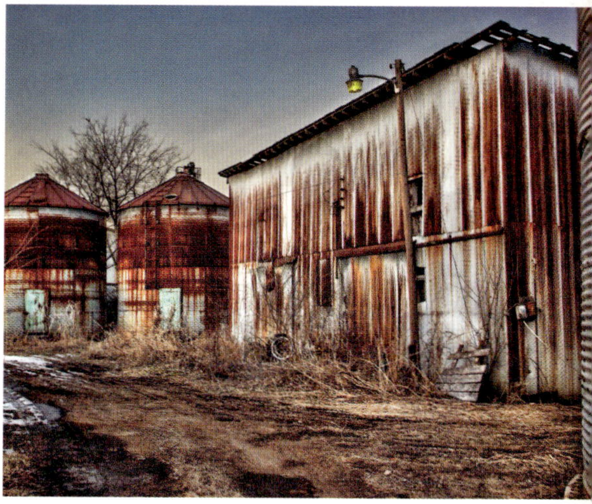

**previous page**—Nebraska.

**top**—New York.

**bottom left and right**—Kentucky.

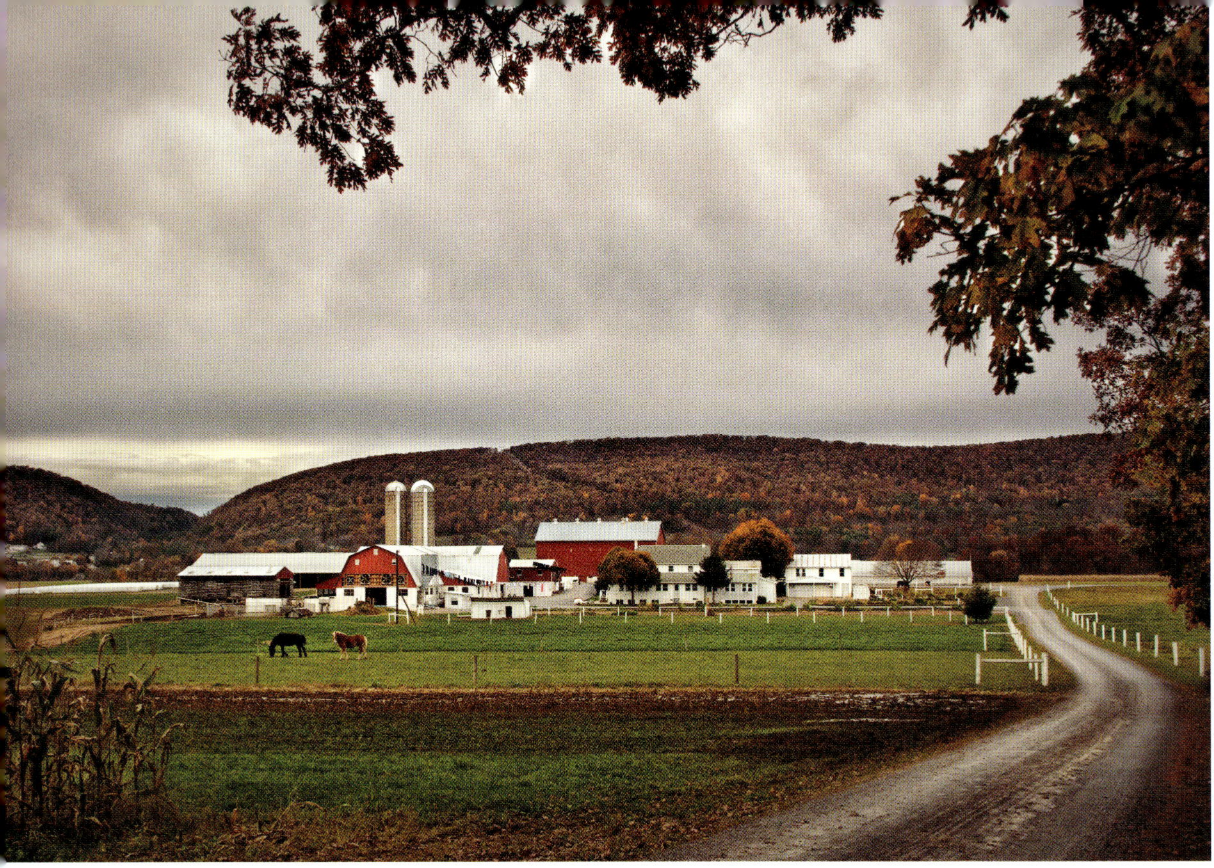

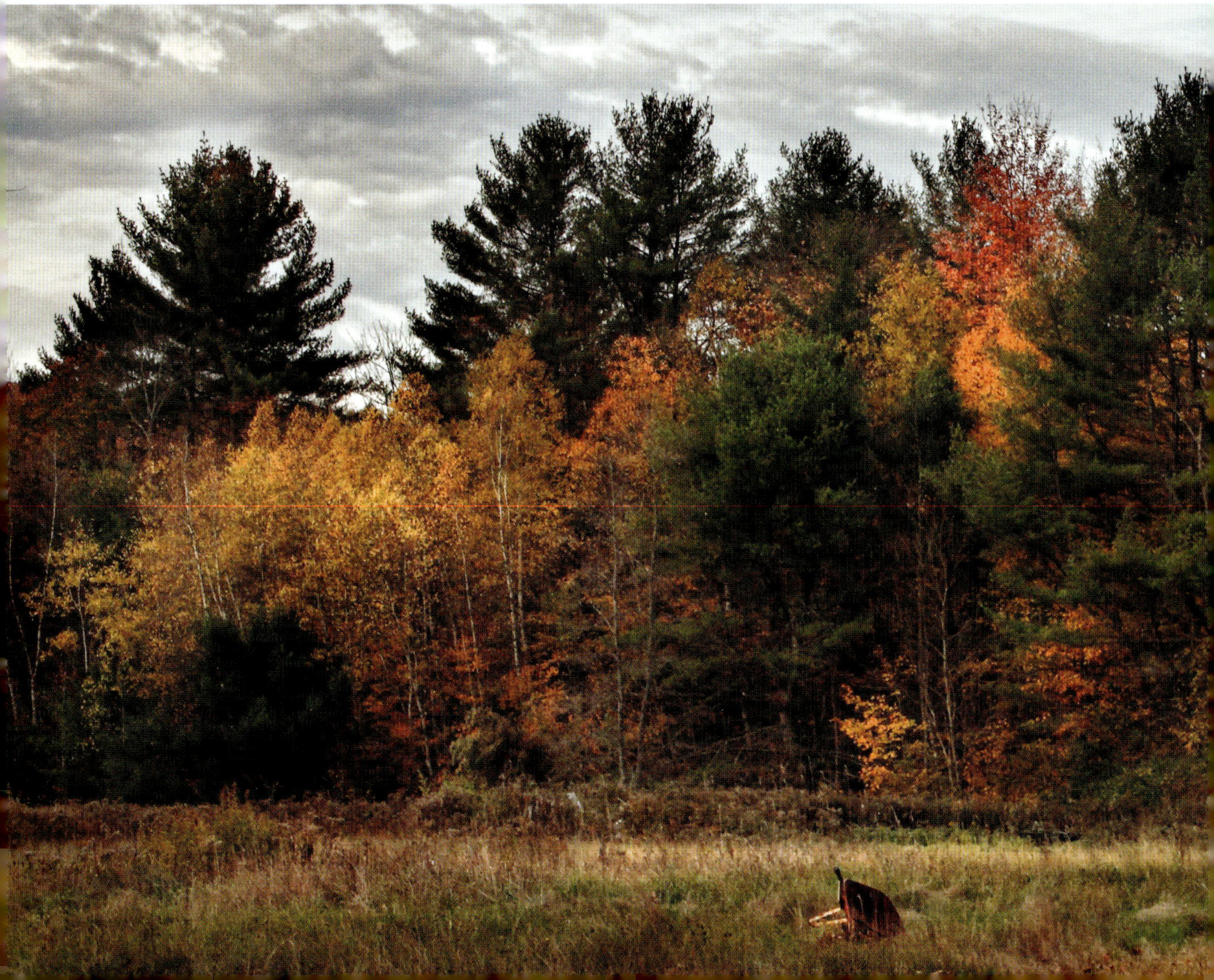

"Putting things together so that they belong together: that is the basis of composition."

—Adolf Fassbender

**previous page, top—**Pennsylvania.

**below—**New Hampshire.

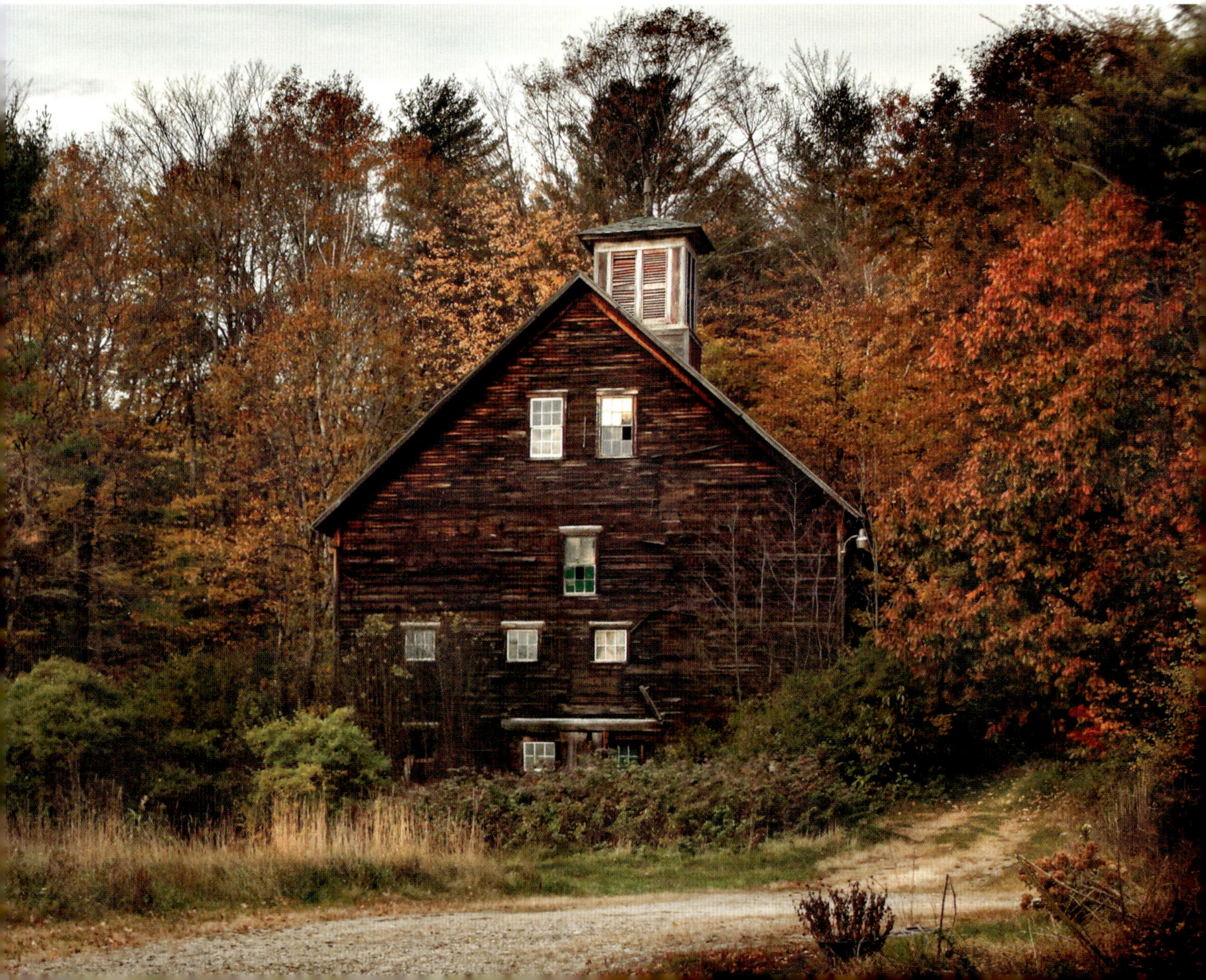

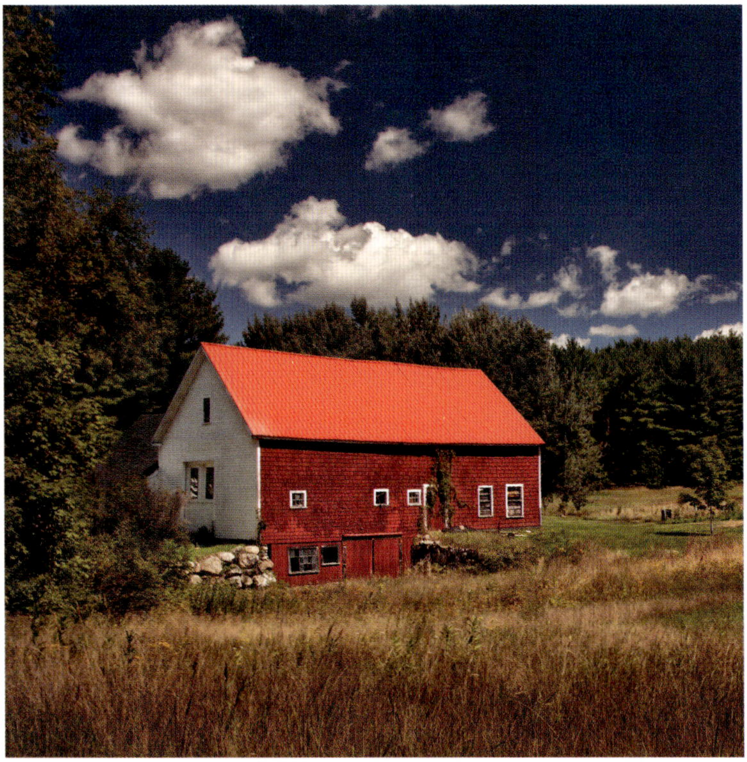

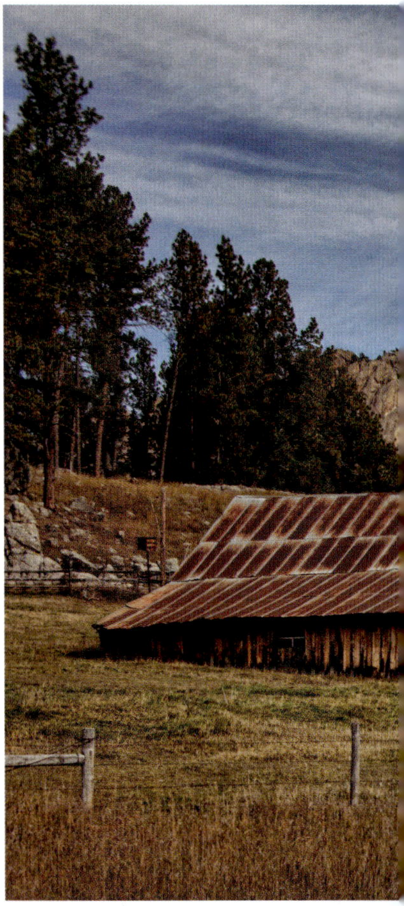

**top left**—New Hampshire.

**top right**—South Dakota.

**bottom left**—New Jersey.

**bottom right**—*Lone Pasture*. Massachusetts.

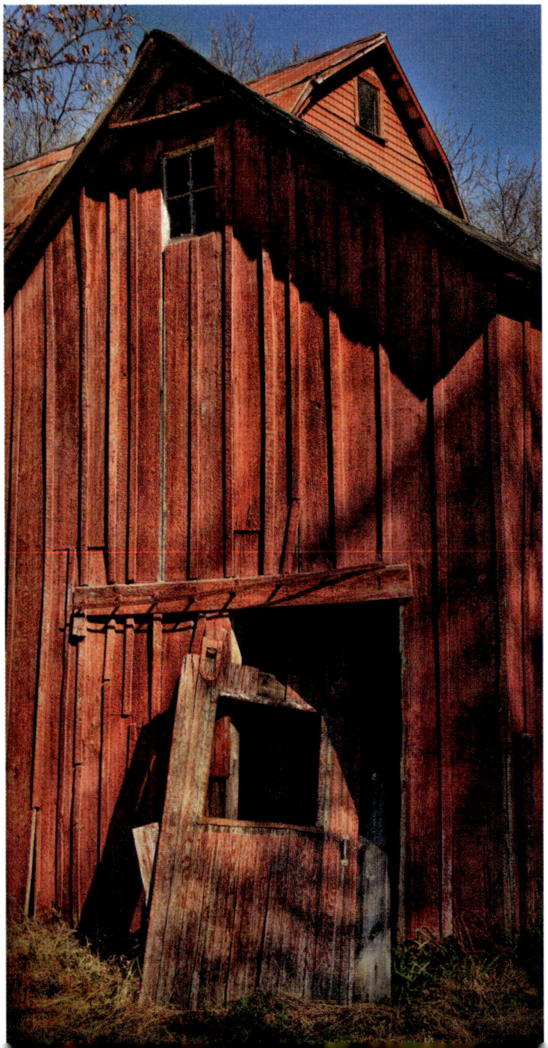

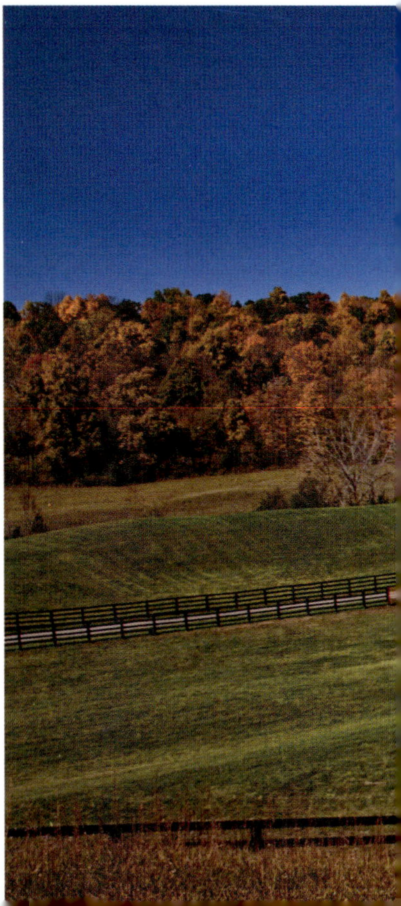

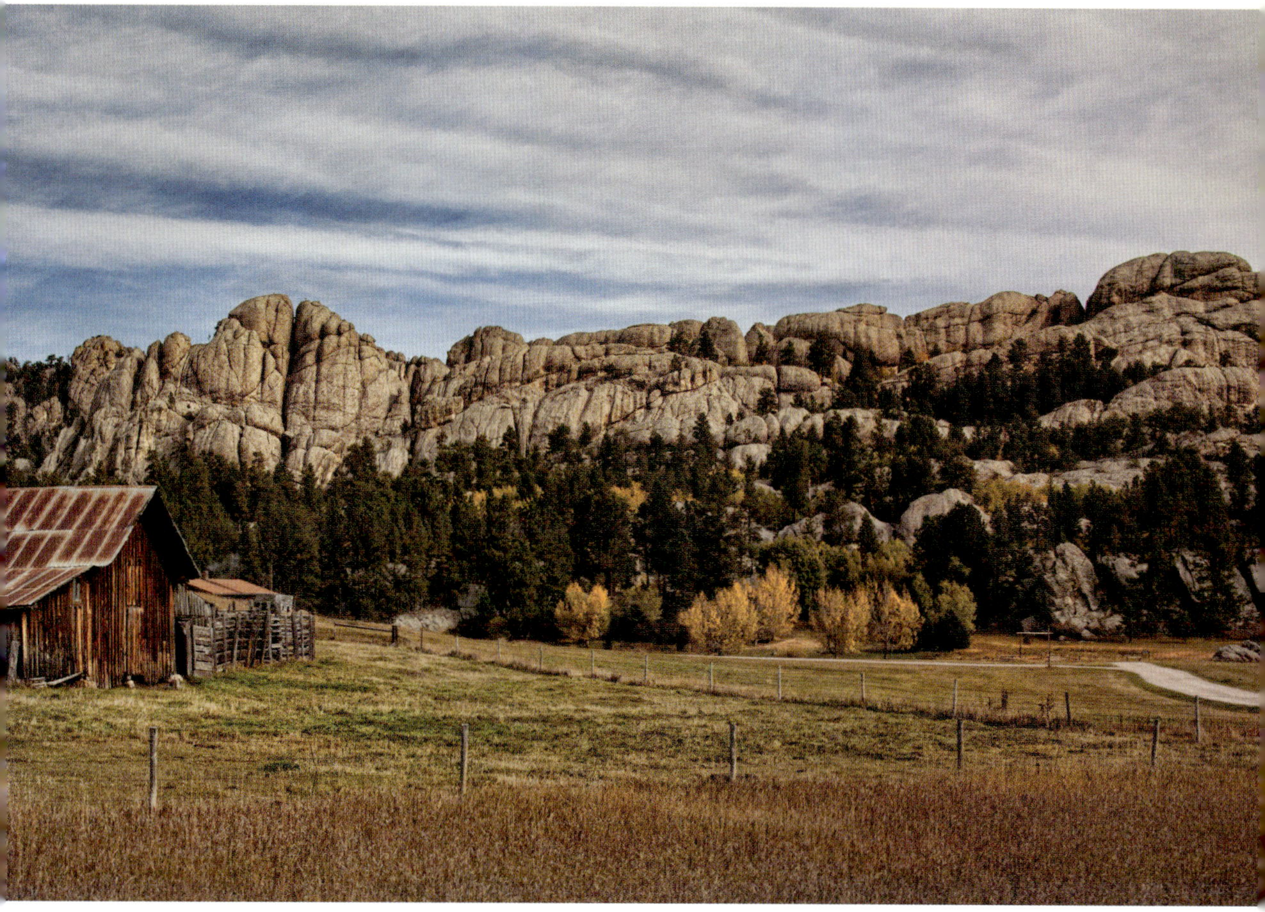

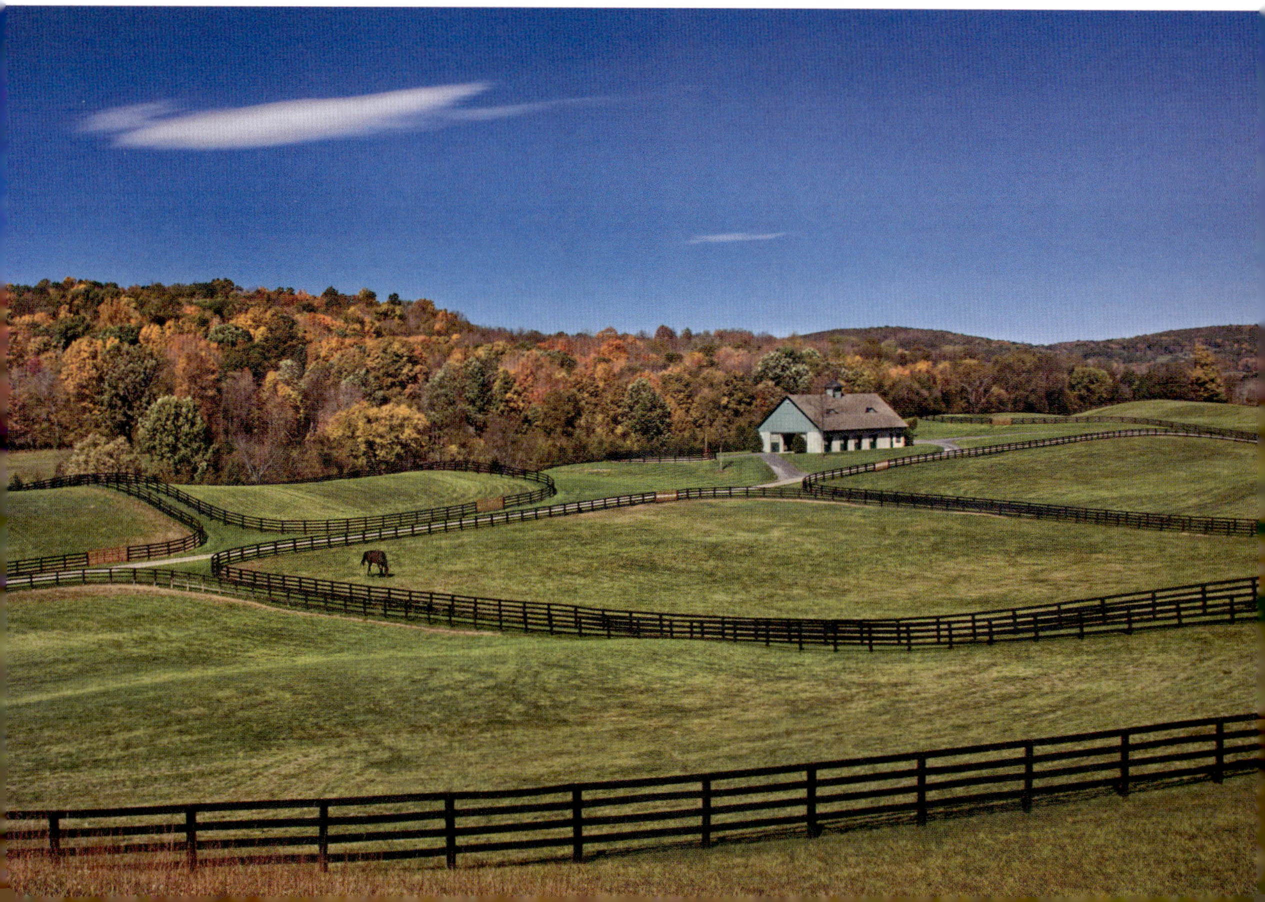

# Decay and Restoration

Across America, many barns have been beaten by the ravages of time. The elements, too, have taken their toll on these once-proud structures. One can't help but feel a sense of loss as these historical treasures fall away.

In this chapter, we'll take a look at some barns that have been forgotten, and others that have been returned to their former glory by devoted owners.

**right**—New York.

**below**—Besieged by fall foliage. Pennsylvania.

**following page**—Restoration needed. Virginia.

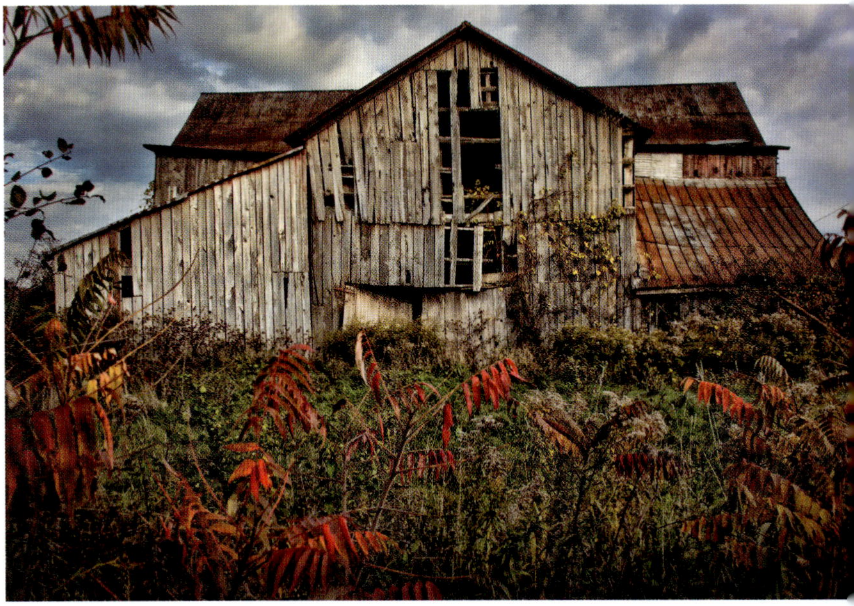

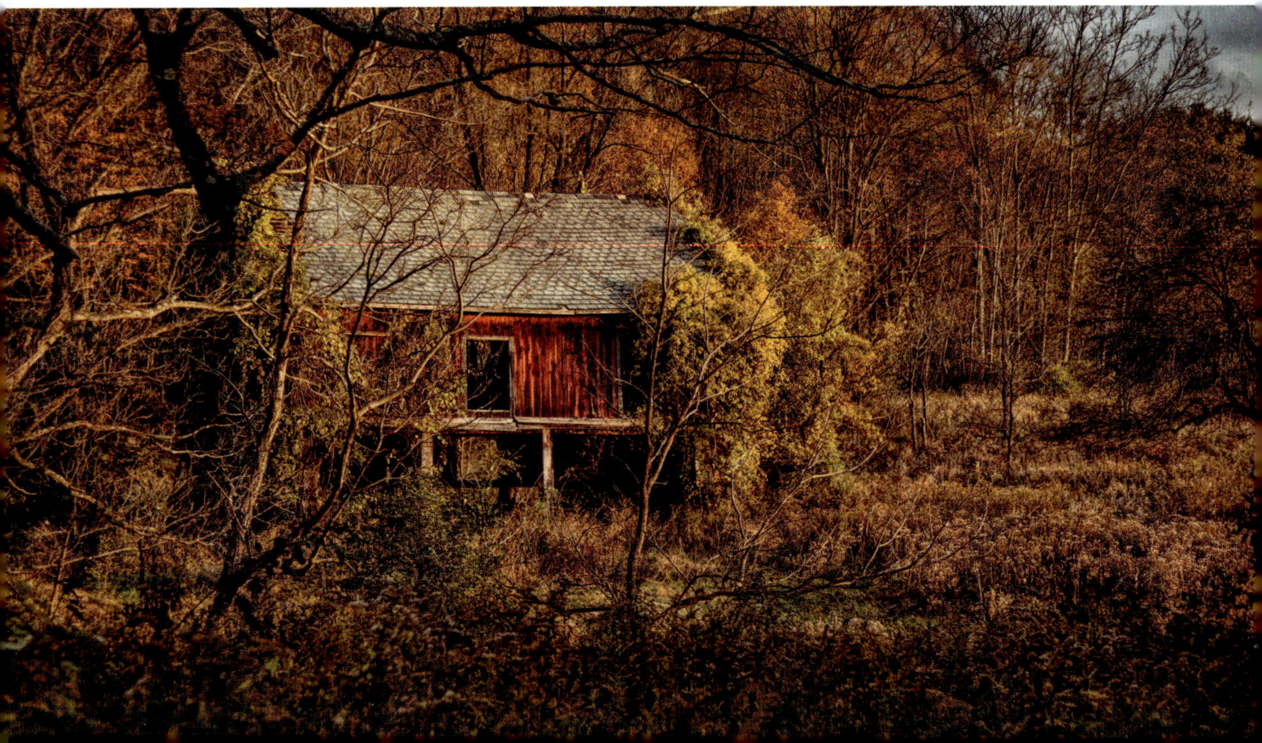

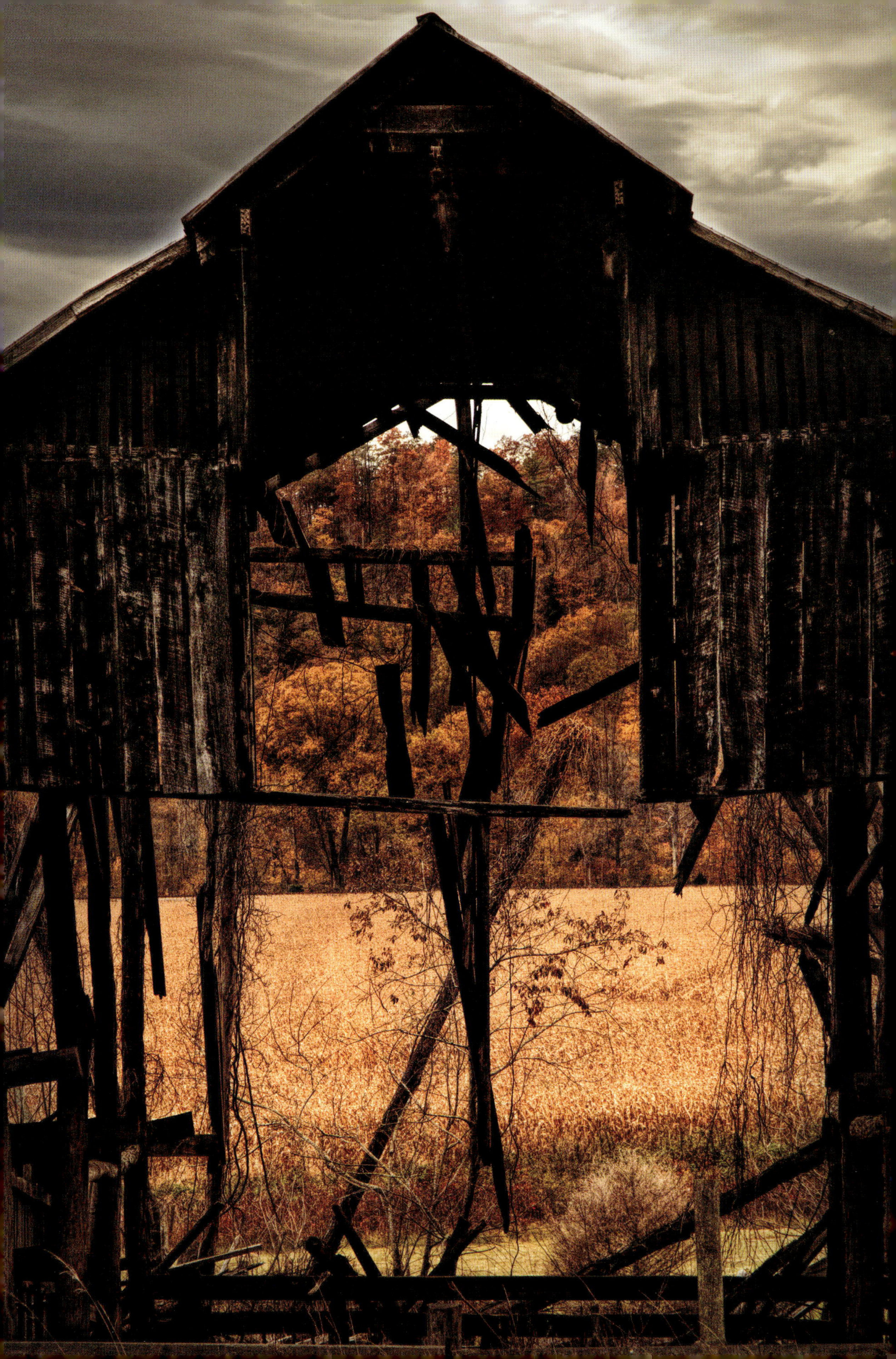

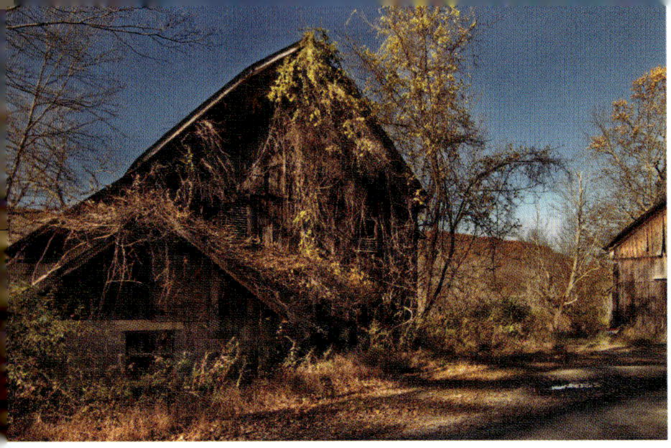

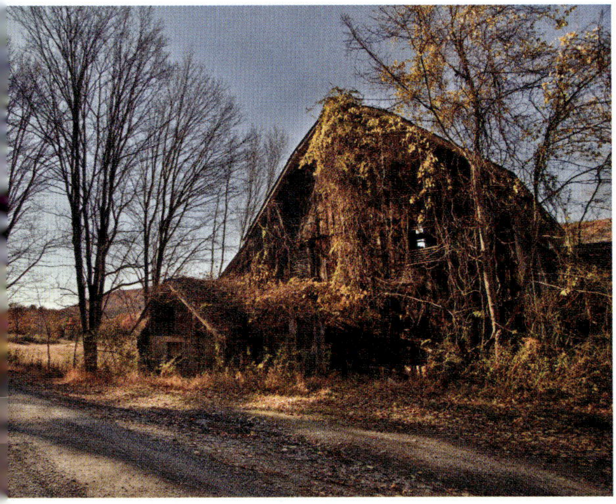

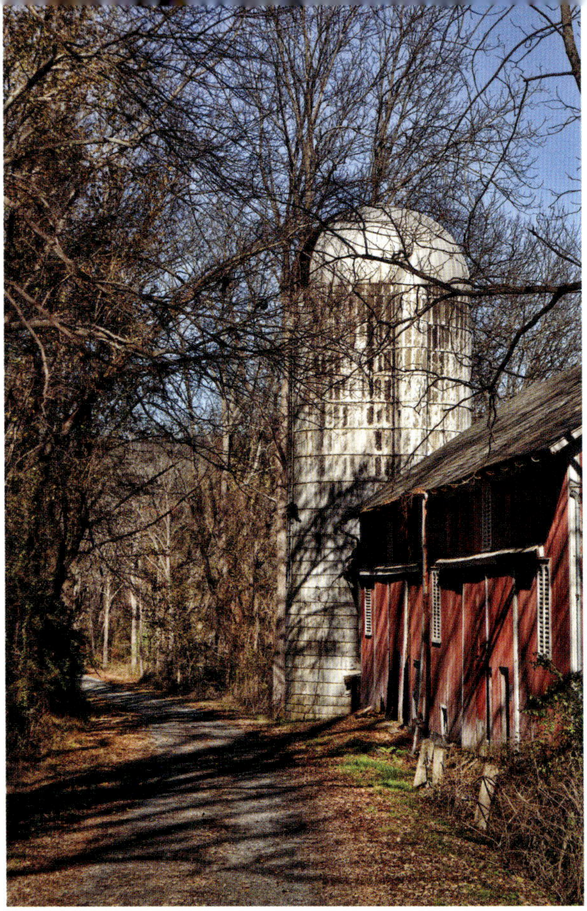

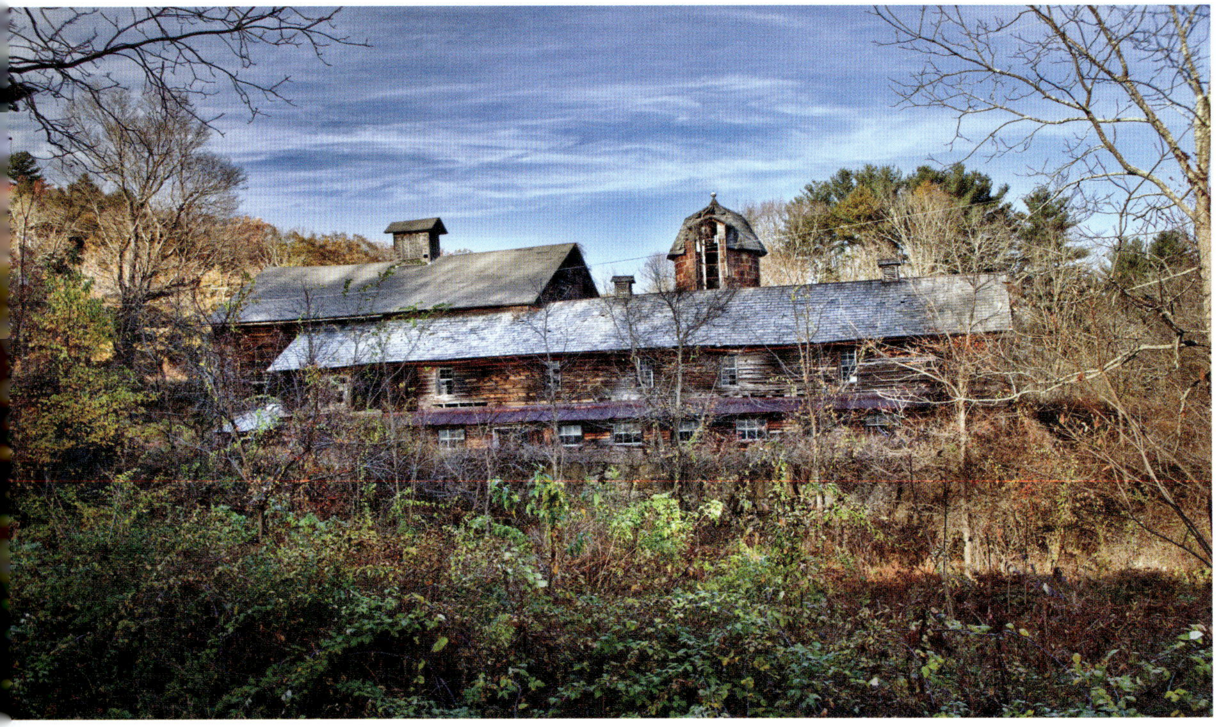

**this page**—These secluded, overgrown barns have fallen into decay in northwestern New Jersey.

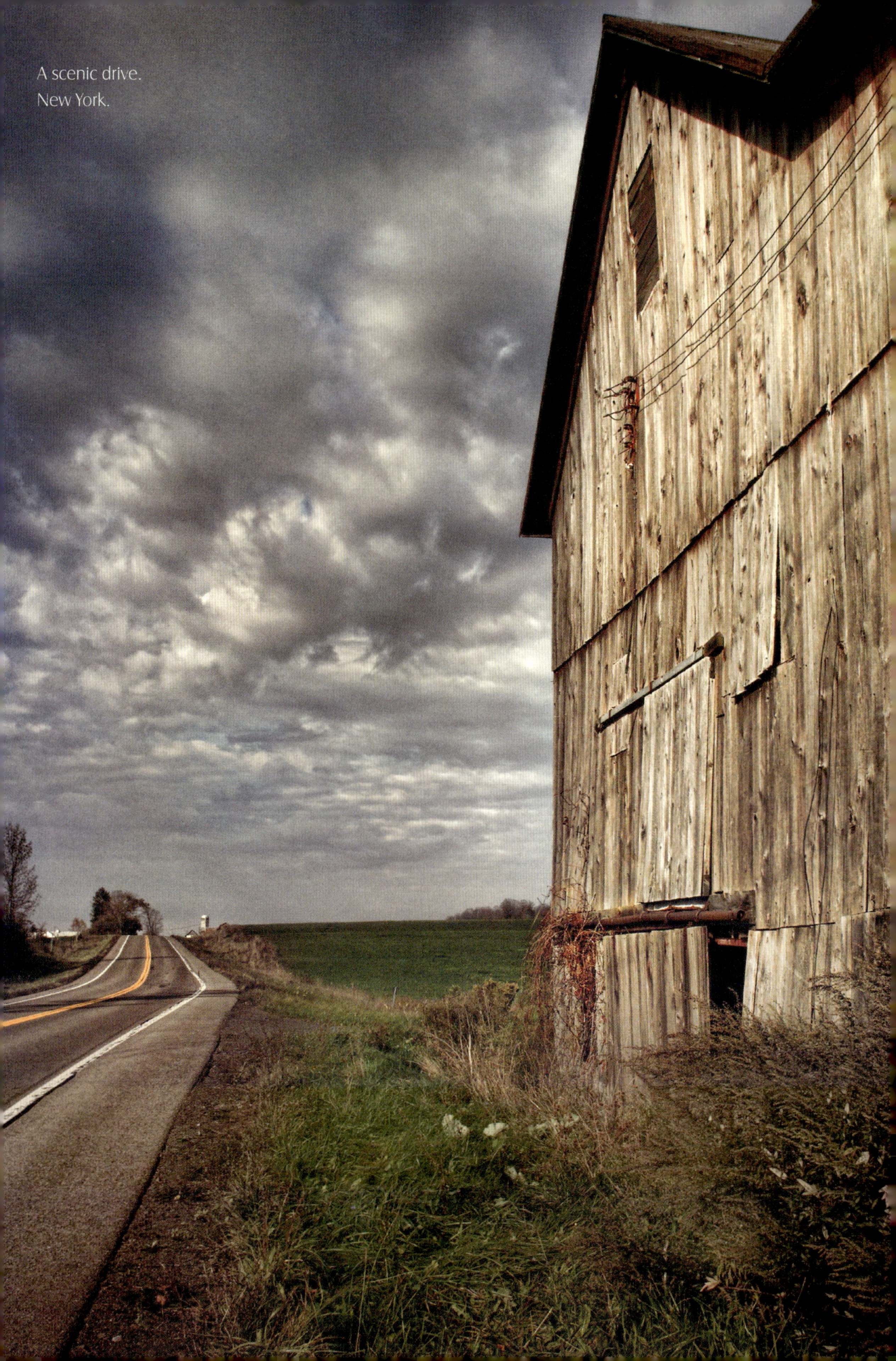

A scenic drive.
New York.

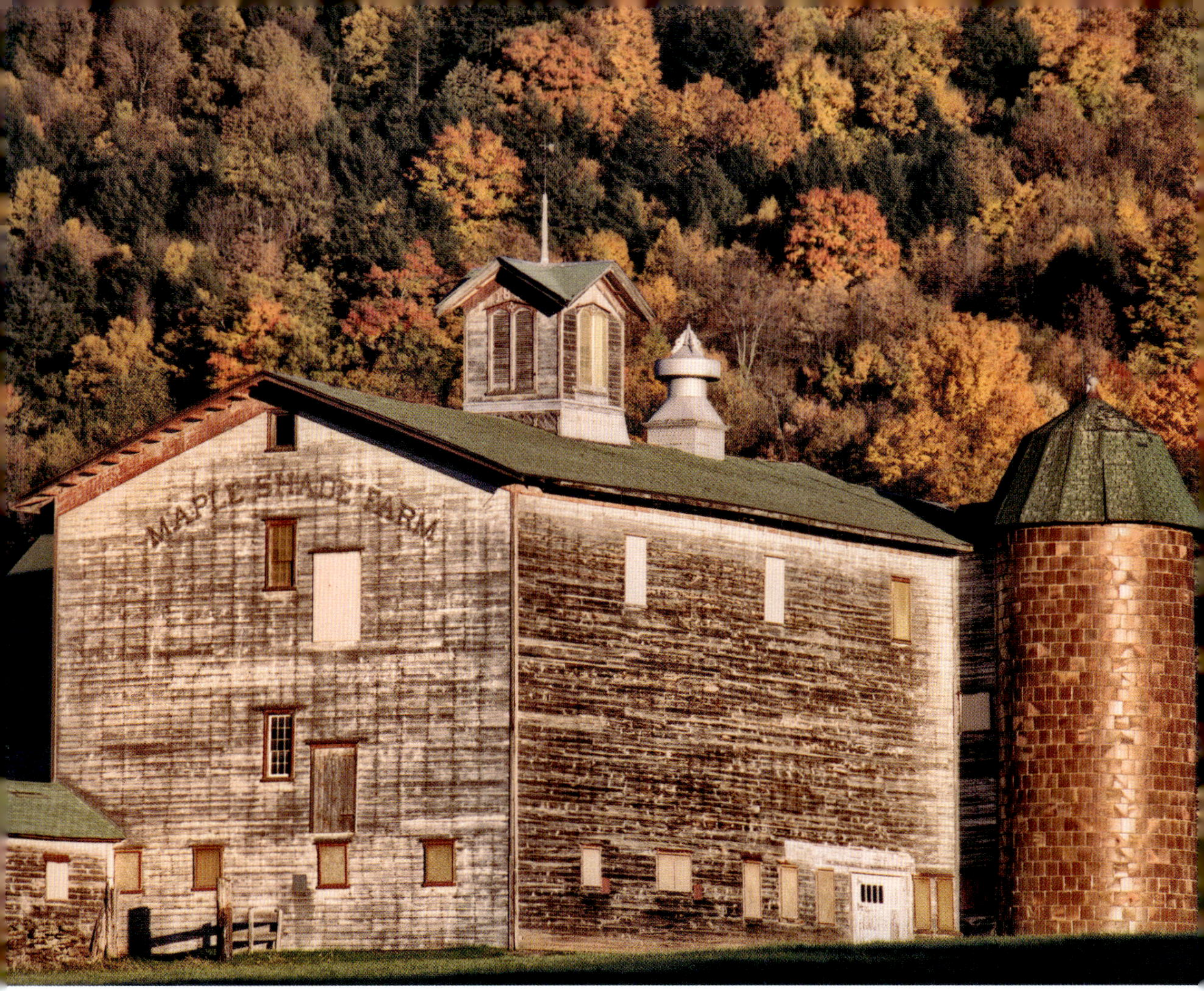
MAPLE SHADE FARM

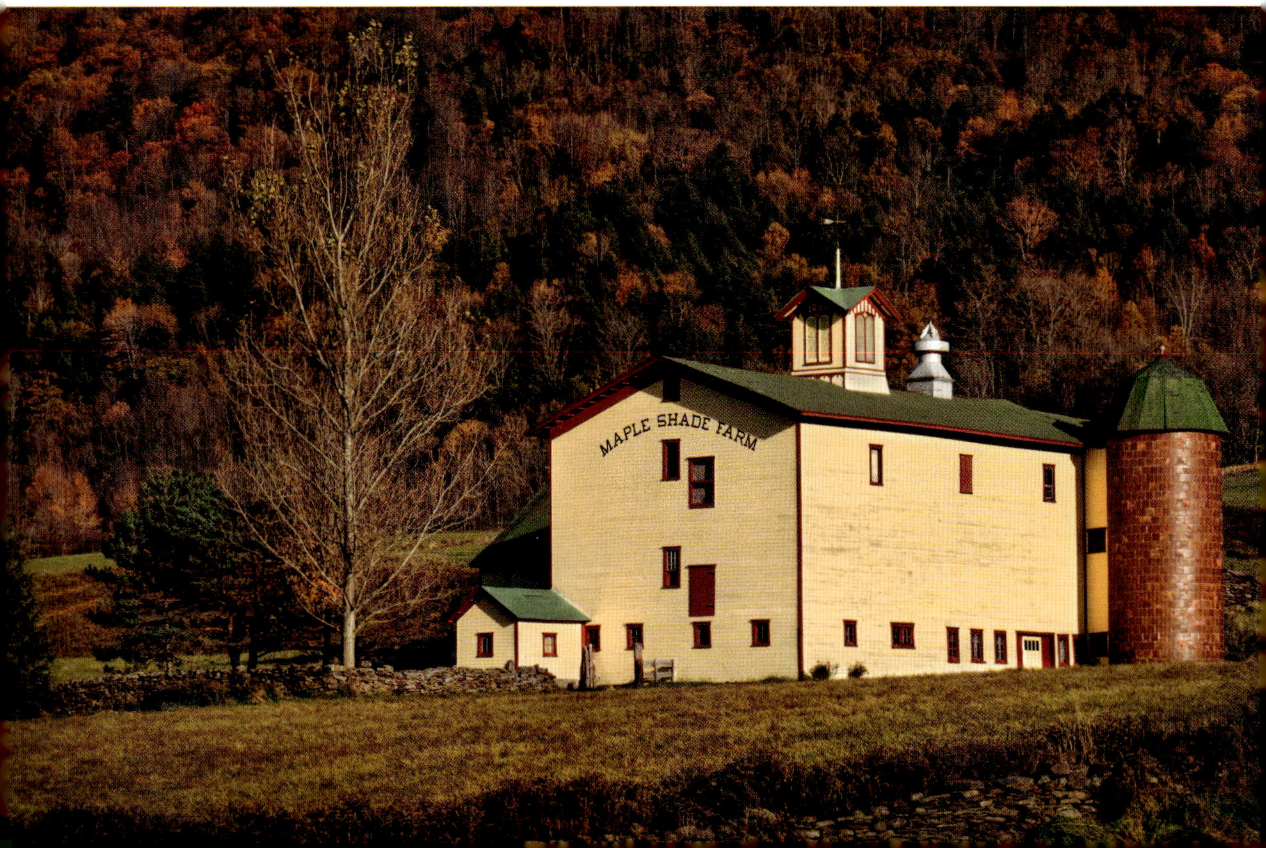
MAPLE SHADE FARM

**previous page, top and bottom**—A great example of restoration. The "before" image (top) was shot on film in 1984.

**top and bottom**—*On Shaky Ground.* This barn is massive in length. It is located in the secluded town of Nahma, Michigan, in the state's upper peninsula. It has fallen down since this film photograph was captured. Little of it remains today.

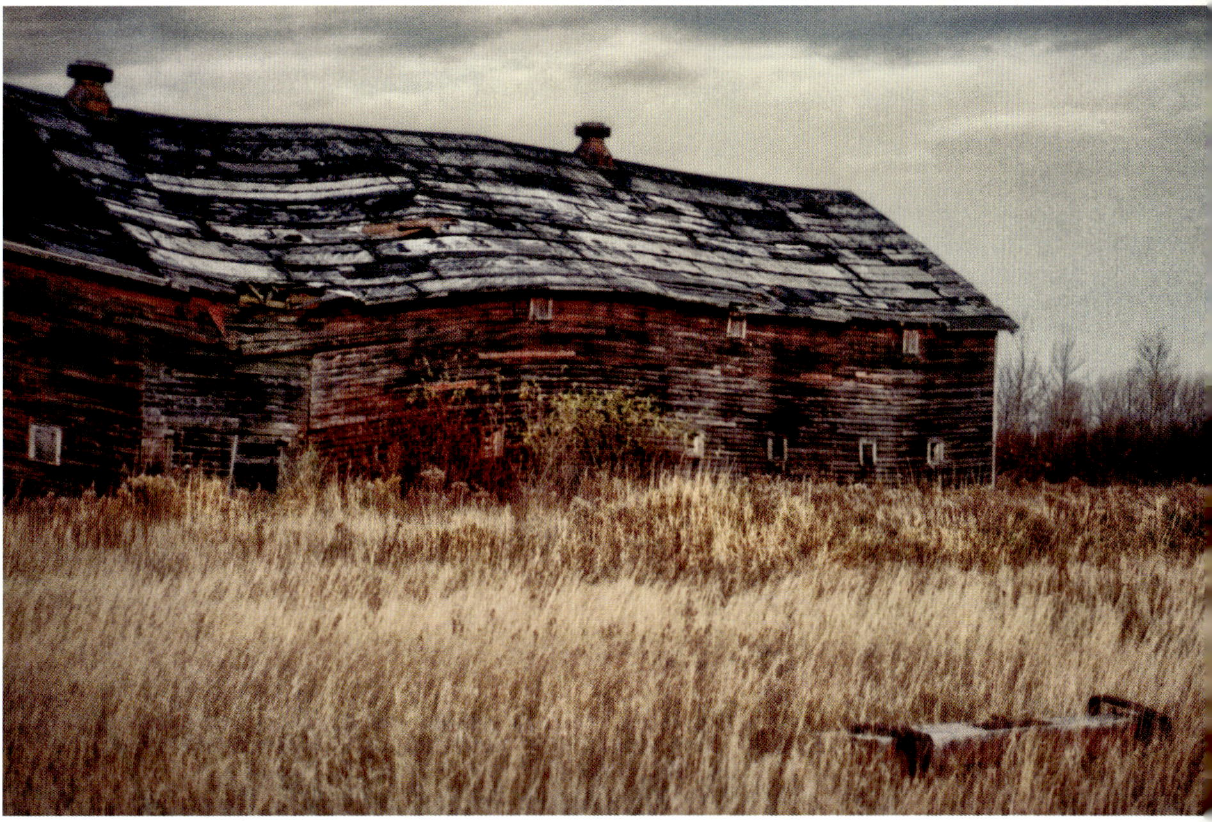

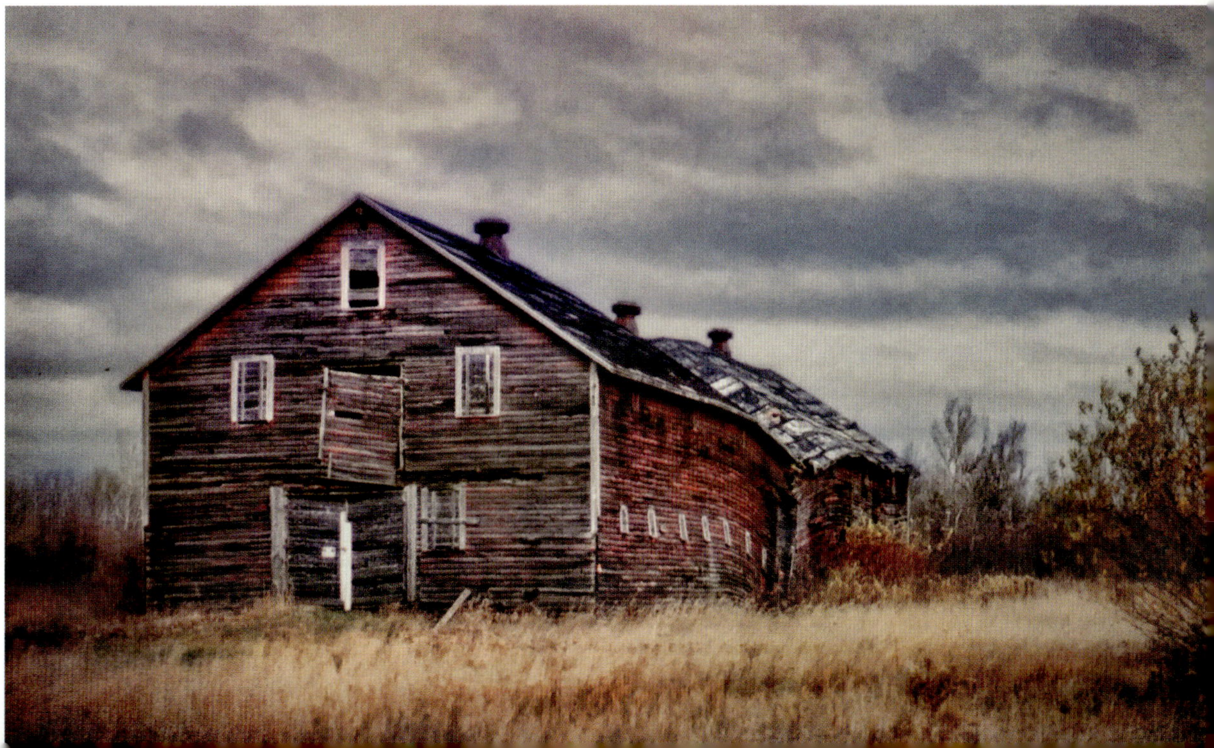

# Conclusion

I hope you enjoyed your journey through the back roads. The next time you are driving down an old country road and see an old barn in the distance, take a moment to think of the families that settled there and built their dream homesteads.

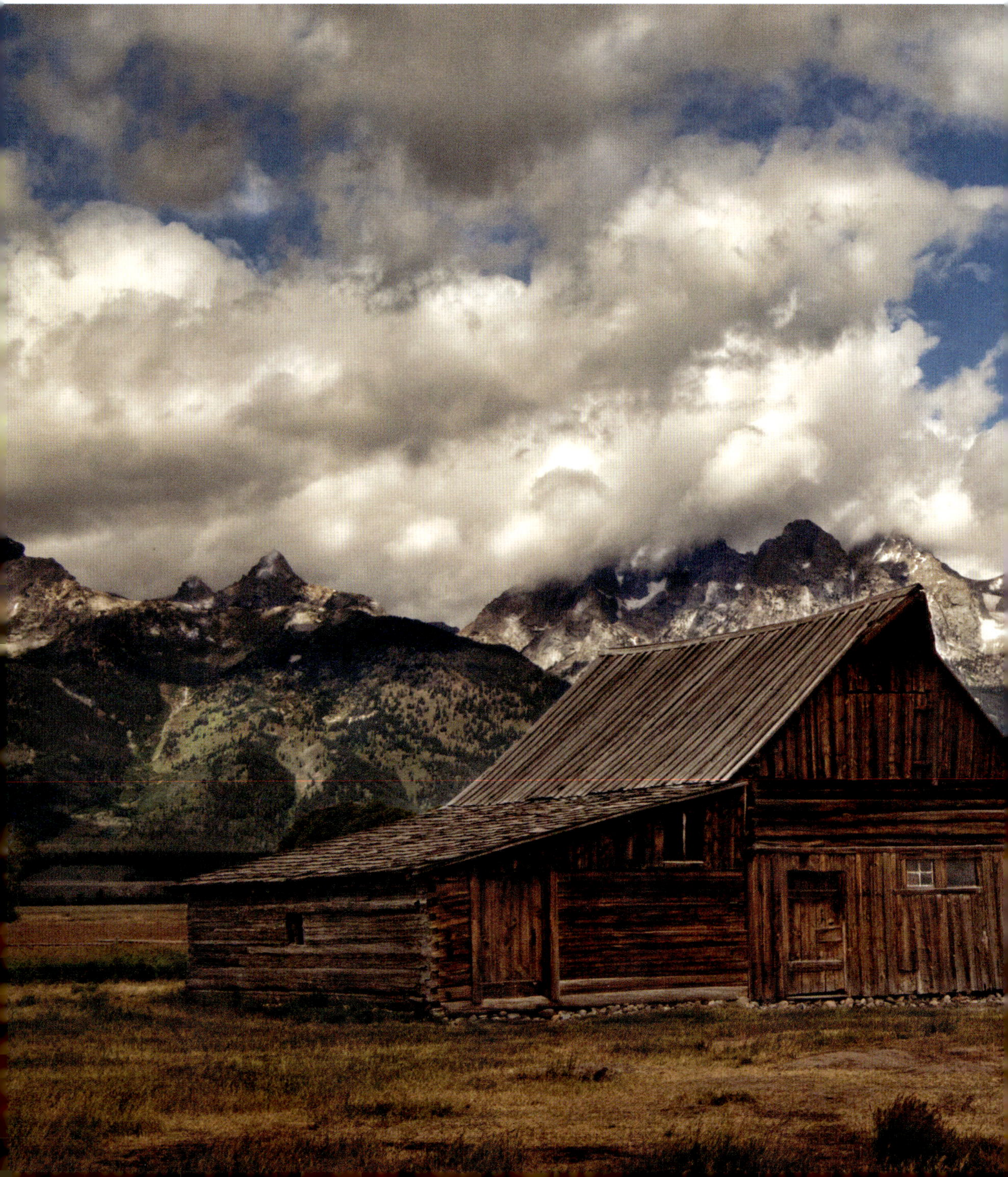

Some of these magnificent structures will live on; others will fall away, resigned to exist only in photographs and in memory.

**below—***Teton Mountain Peace*. Wyoming.

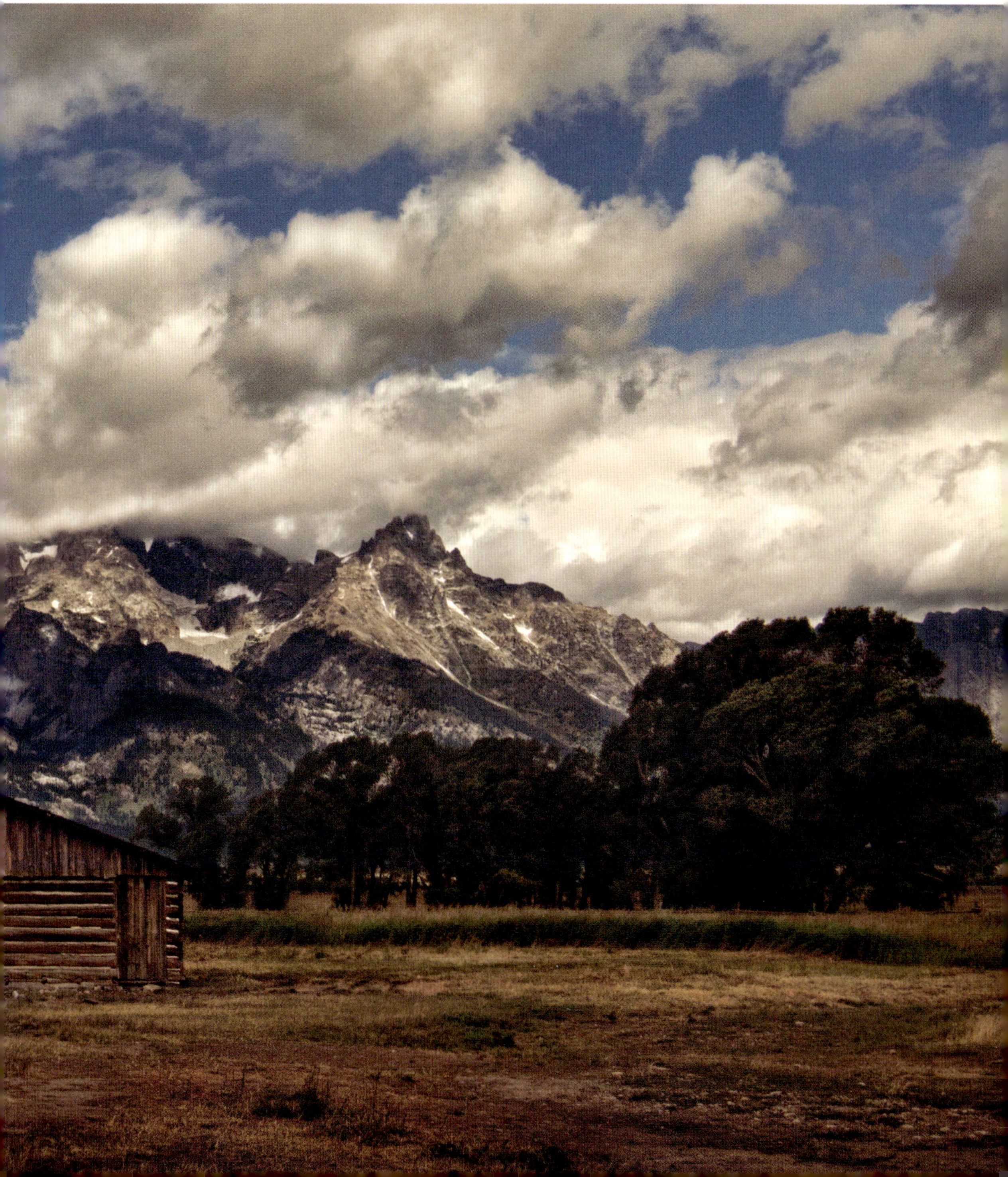

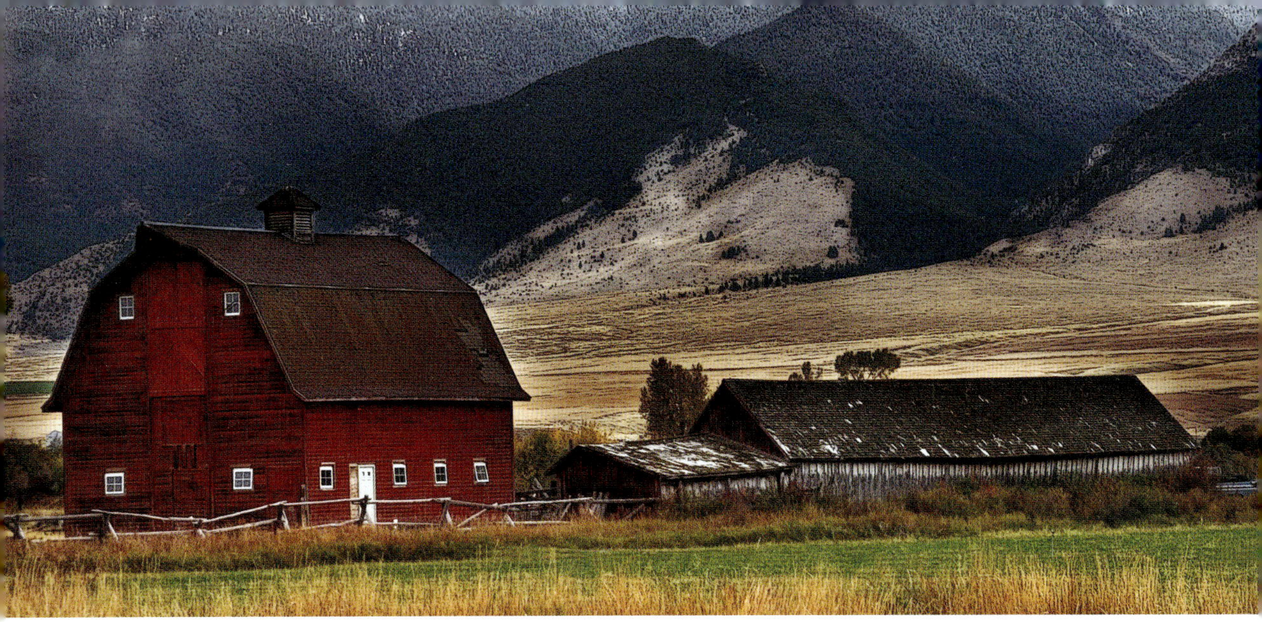

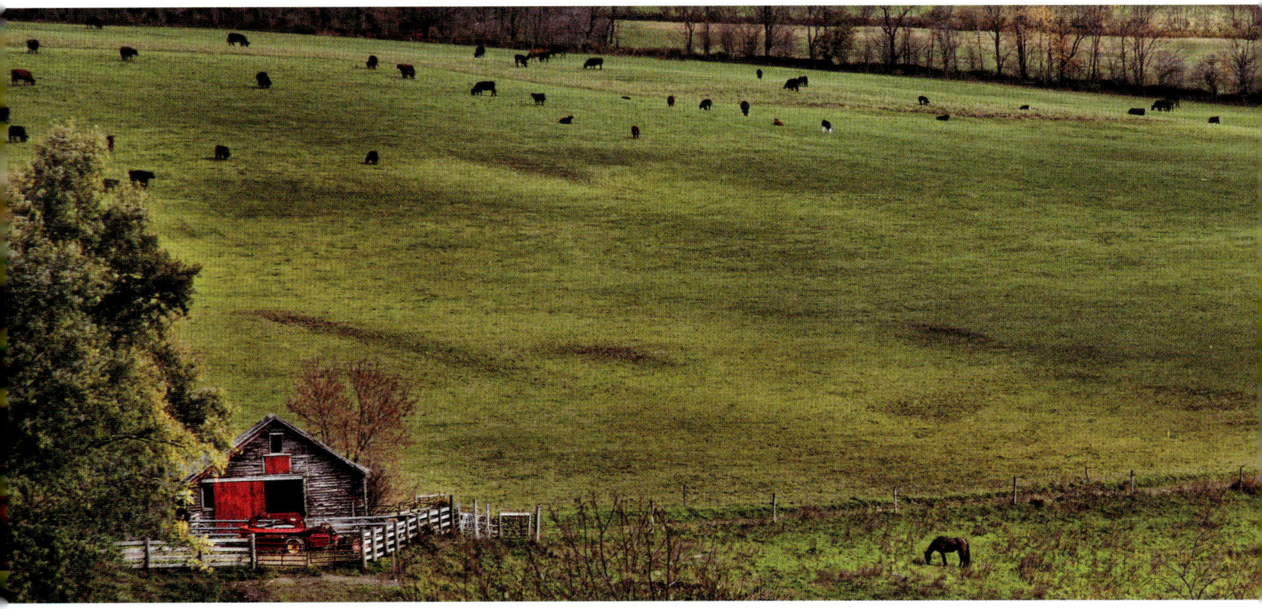

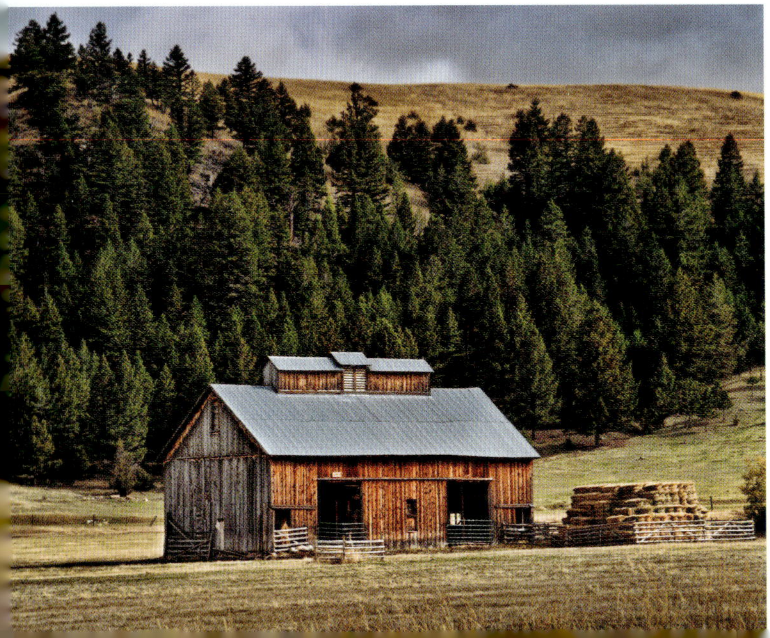

**top**—Salmon, Idaho.

**center**—Massachusetts.

**bottom**—Montana.

# Index

# AmherstMedia.com

- *New books every month*
- *Books on all photography subjects and specialties*
- *Learn from leading experts in every field*
- *Buy with Amazon (amazon.com), Barnes & Noble (barnesandnoble.com), and Indiebound (indiebound.com)*
- *Follow us on social media at: facebook.com/AmherstMediaInc, twitter.com/AmherstMedia, or www.instagram.com/amherstmediaphotobooks*

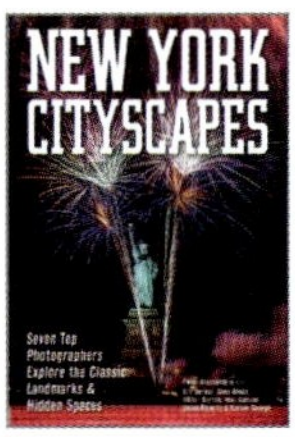


**New York Cityscapes**
SEVEN TOP PHOTOGRAPHERS EXPLORE THE CLASSIC LANDMARKS & HIDDEN SPACES

Native New Yorkers explore the city's architecture and neighborhoods. *$24.95 list, 7x10, 128p, 200 color images, index, order no. 2201.*

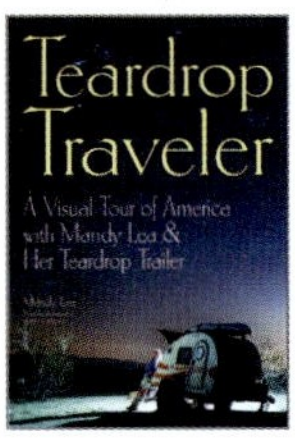


**Teardrop Traveler**

Photographer Mandy Lea left her job for life on the road—and along the way, captured both life lessons and incredible images of America. *$24.95 list, 7x10, 128p, 200 color images, index, order no. 2187.*

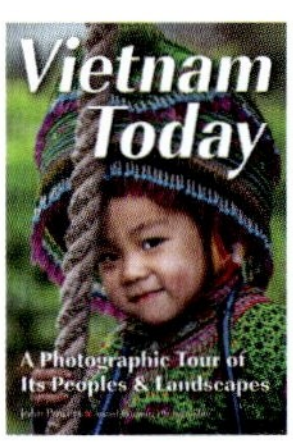


**Vietnam Today**
A PHOTOGRAPHIC TOUR

Vietnam vet and acclaimed travel photographer John Powers explores the rural and urban sides of Vietnam today. *$29.95 list, 7x10, 128p, 180 color images, index, order no. 2207.*

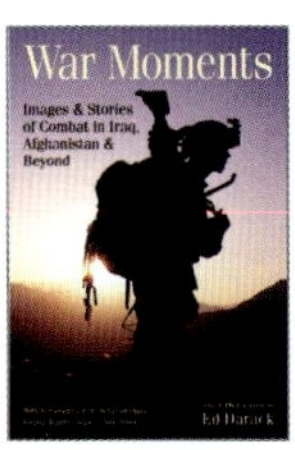


**War Moments**

Ed Darack's captivating images of front-line combat and deeply engaging stories bring modern combat to life. *$24.95 list, 7x10, 128p, 118 color images, index, order no. 2204.*

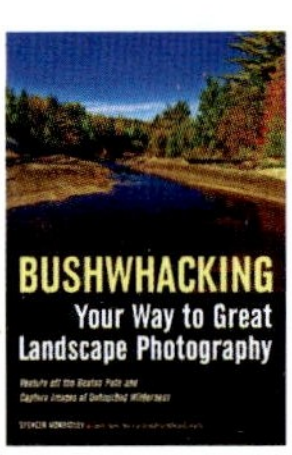


**Bushwhacking:** Your Way to Great Landscape Photography

Spencer Morrissey presents his favorite images and the back-country techniques used to get them. *$37.95 list, 7x10, 128p, 180 color images, index, order no. 2111.*

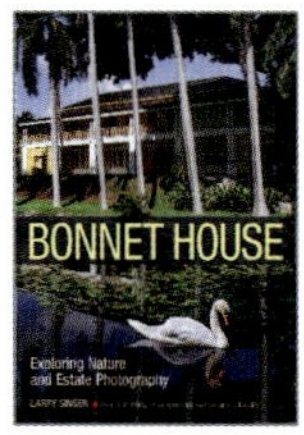


**Bonnet House** EXPLORING NATURE AND ESTATE PHOTOGRAPHY

Photograph the grounds of a historic Florida estate—and the plants and animals that make their home there. *$24.95 list, 7x10, 128p, 180 color images, index, order no. 2153.*

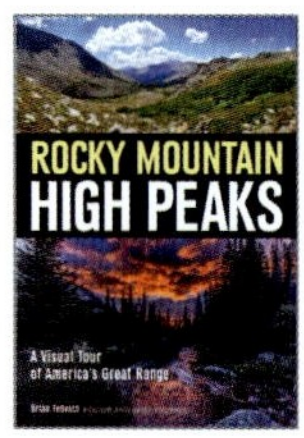


**Rocky Mountain High Peaks**

Explore the incredible beauty of America's great range with Brian Tedesco and a team of top nature photographers. *$24.95 list, 7x10, 128p, 180 color images, index, order no. 2154.*

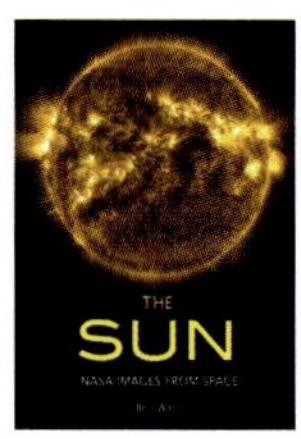


**The Sun**
IMAGES FROM SPACE

Travel to the center of our solar system and explore the awesome beauty of the star that fuels life on earth. *$24.95 list, 7x10, 128p, 180 color images, index, order no. 2180.r images, index, order no. 2217.*

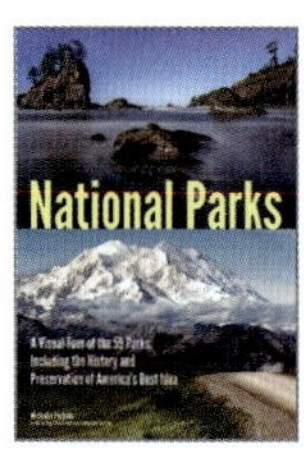


**National Parks**

Take a visual tour through all 59 of America's National Parks, exploring the incredible histories, habitats, and creatures these lands preserve. *$24.95 list, 7x10, 128p, 375 color images, index, order no. 2193.*

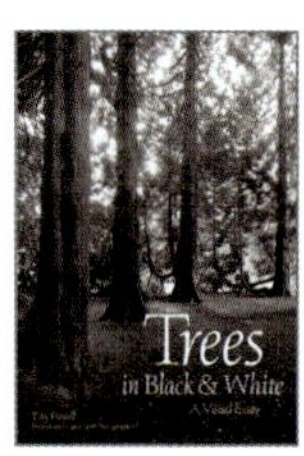


**Trees in Black & White**

Follow acclaimed landscape photographer Tony Howell around the world in search of his favorite photo subject: beautiful trees of all shapes and sizes. *$24.95 list, 7x10, 128p, 180 images, index, order no. 2181.*